A Slave Between Empires

A SLAVE BETWEEN EMPIRES

A Transimperial History of North Africa

M'HAMED OUALDI

COLUMBIA UNIVERSITY PRESS *NEW YORK*

Columbia University Press
Publishers Since 1893
New York   Chichester, West Sussex
cup.columbia.edu

Library of Congress Cataloging-in-Publication Data

Names: Oualdi, M'hamed, 1974– author.
Title: A slave between empires : a transimperial
history of North Africa / M'hamed Oualdi.
Description: New York : Columbia University Press, [2020]. |
Includes bibliographical references and index.
Identifiers: LCCN 2019026953 | ISBN 9780231191869 (cloth) |
ISBN 9780231549554 (e-book)
Subjects: LCSH: Ḥusayn, Jinirāl, 1828–1887—Estate. |
Africa, North—History—1517–1882. | Africa, North—History—1882– |
Tunisia—Officials and employees—Biography. | Tunisia—Relations—Europe. |
Europe—Relations—Tunisia. | Africa, North—Colonization. |
Africa, North—Civilization—Turkish influences. |
Mediterranean Region—Civilization.
Classification: LCC DT204 .O93 2020 | DDC 961—dc23
LC record available at https://lccn.loc.gov/2019026953

Cover design: Milenda Nan Ok Lee
Cover image: Private Collection. Courtesy of Habib Bouhageb

To Hassen, Rim, Wassil, Farah, Sheima, and Zohra

Contents

Acknowledgments

AFTER EIGHT years spent trying to understand General Husayn and his vanishing Ottoman-Tunisian world, it is my great pleasure to thank the institutions, colleagues, and friends who have been instrumental in the realization of this book.

For their tremendous support, I am indebted to the following institutions: Princeton University's Department of Near Eastern Studies and Department of History, the William Hallam Tuck Memorial Fund Summer Research and Travel Grant, and the Learned Travel Society Fund; the Department of Arabic Studies at INALCO (Paris); the European University Institute in Florence, where this research project was born in 2009 and where the first three chapters were written in 2015 thanks to a Braudel Fellowship; the "Ecole Française" in Rome, which has been a safe haven for writing the three last chapters; and the Institute of Advanced Studies in Paris.

I also want to express my deep gratitude to the archivists and librarians who facilitated access to primary sources at the Archives of the French Ministry of Foreign Affairs in La Courneuve and Nantes, the French National Archives in Aix-en-Provence, the Public Record Office in London, the National Archives and Records Administration in Washington, the Archives of the Italian Ministry of Foreign Affairs in Rome, the State Archives in Florence and Lucca, the Tunisian National Library, and the Tunisian National Archives, where I could always count on the help and expertise of my friend and colleague Mabrouk Jebahi. I am also grateful to Youssef Ben Smail for sharing a file from the Ottoman consul in Florence that he found at Başbakanlık Osmanlı Arşivi in Istanbul. Habib Bouhageb provided invaluable documents, information, and pictures from his private collection, including the fascinating picture used for this book cover: a portrait of Husayn taken by the famous Parisian photographer Felix Nadar.

During a five-hour defense and workshop in December 2016, the first version of this manuscript, written in French, greatly benefited from the critical insights and generous comments of Fatma Ben Slimane (Tunis I University), Olivier Bouquet (Paris-Diderot University), Catherine Brice (Paris Est Créteil-Val-de-Marne University), Simona Cerutti (EHESS-Paris), James McDougall (Oxford University), and Jocelyne Dakhlia (EHESS-Paris), to whom I owe particular thanks for her tremendous help, encouragement, and the thoughtful comments she tirelessly provided, page after page and chapter after chapter.

Adrian Morfee translated the first version of the manuscript into English during summer 2017. Some sections were presented at various conferences and venues, where Etty Terem (Rhodes College), Pascal Menoret (Brandeis University), Jane Hathaway (Ohio State University), Fahad Ahmad Bishara (University of Virginia), and Emmanuelle Saada (Columbia University) provided constructive and engaging feedback. Particular thanks are owed to Michael Laffan, Molly Green, Linda Colley, Muhammad Qasim Zaman, Sükrü Hanioglu, Michael Reynolds, William Chester Jordan, Max Weiss, Cyrus Schayegh, Phil Nord, and all of my colleagues in the departments of History and Near Eastern Studies at Princeton University, who shared their questions and comments after presentations of this project or after having read chapters of the translated manuscript.

I am grateful to my editor at Columbia University Press, Caelyn Cobb, for having understood the project, its use of microhistory, and its intervention in the historiography of North Africa. Monique Briones, Susan Pensak, and Ben Kolstad were also instrumental in the production of the book. My heartfelt appreciation goes to the two anonymous reviewers who provided invaluable advice for revising, restructuring, and rewriting the manuscript. At this stage, S. C. Kaplan was also instrumental in copyediting the revised English version of the manuscript. Troy Tice and Jakub Novák read the revised version. As always, my colleague Nilüfer Hatemi has been more than generous with her time, translating three documents from Ottoman Turkish into English for me.

To complete this project, over the years, in North America, North Africa, and Europe, I was able to count on the support of students, colleagues, close friends, and thoughtful people.

At Princeton, since 2013, these include Muriel Cohen and Emmanuel Szurek, Eugenia Palieraki and Angelos Dalachanis, Jessica Metcalf and Sean McMahon, Elsa Devienne, Julien Ayroles and Sarah Kocher, Samia Henni and Pascal Schwaighofer, Miriam Lowi and Abdallah Hammoudi, Sadaf Jaffer and Dan Sheffield, Lara Harb and Fulvio Domini, Giulia Puma and Guillaume Calafat, Jonathan Gribetz, the admirable Lucette Valensi and Abraham Udovitch, Jakub Novák, and my dearest friend Nadia Benabid, all of whom reminded me that there is a life outside of campus and academia, inviting me and my partner to their warm places for lavish dinners and lively discussions. I am also more than indebted to the efficient staff of our departments, especially to Bill Blair, who has read and edited my papers and chapters. I have learned from many of the students who enrolled in my classes, particularly from Yasmina Aidi, Edna Bonhomme, Brahim Guebli, Peter Kitlas, Joshua Picard, Matthew Schumann, and Julian Weideman, with whom I have worked closely on their fascinating projects in recent years.

In Tunis, I had the chance to discuss Husayn's legacies with Anne-Marie Planel and Sadok Boubaker, Kmar Bendana and Khaled Kchir, Fatma Ben Slimane, Sami Bargaoui, and Isabelle Grangaud, who has been hugely influential in reshaping my understanding of nineteenth-century North Africa, the issue of archives, and the notion of "context." On the other side of the Mediterranean, Muriam Davis, Thomas Serres, Nadia Marzouki, Marie, Luca, Valentina, Sasa, and Ariana illuminated my stays during two semesters in Florence and Rome. For this research project, I learned a great deal from Catherine Mayeur-Jaouen, Julia Clancy-Smith and Amy Kallander.

I have enjoyed every minute of my discussions with my colleagues and friends Camille Lefebvre and Natividad Planas, and my friend Marie Fradet, who was kind enough to quickly edit the French manuscript. But in each and every one of these places, nothing would have been the same without the understanding and support of Augustin, to whom I dedicate this book.

Note on Transliteration

TO TRANSLITERATE the Arabic and some Ottoman Turkish terms, I have followed a simplified version of the system used in the *International Journal of Middle East Studies*. I have used a simplified version of transliteration for the proper nouns (family names, names of locations . . .) but I have used the IJMES system for terms in Arabic and Ottoman Turkish. When historical characters, places, or institutions are well known (such as the Tunisian president Habib Bourguiba), I have used the common English spelling.

A Slave Between Empires

Introduction

A North African Land and Its Ottoman and Colonial Legacies

IN 1881, the subjects and dignitaries of the former Ottoman province of Tunis were deeply affected when the French occupied their country, putting an end to three centuries of Ottoman rule. French colonial rule weighed heavily on Tunisians' minds. Even their bodies seemed to register the blow. Far away, in the Italian city of Florence, the health of former Tunisian dignitary Husayn Ibn 'Abdallah (figure 0.1) deteriorated. The more Husayn rooted himself in Florence and the more he distanced himself from Tunis, the more physical pain crippled the sixty-year-old man. During the winter of 1882, shortly after the French conquest of Tunis, Husayn came down with a fever and remained in his bed for "almost 2 entire months." Rumors spread across Tunis, insinuating that Husayn had passed away. While convalescing, Husayn contracted whooping cough before deciding that a change of "air in Naples, on Vesuvius" would do him good.[1] A little less than four years later, he was no longer able to descend the stairs of his apartment in Florence, and he seemed to be suffering from heart disease.[2] The following year, in his final spring, Husayn found himself successively unable to read and speak. He lost hearing in one ear and could only walk with a cane.[3] In the final days of his life, three nurses and a doctor watched over him day and night.[4]

The physical and mental traumas of colonization endured by North Africans such as Husayn were so profound that historians of North Africa have primarily concentrated their studies on colonial shock, not just in

FIGURE 0.1 Portrait of General Husayn Ibn 'Abdallah taken by Parisian photographer Nadar

Tunisia but in several other countries in this region, also called the *Maghreb*, or "western part of the Muslim lands" in Arabic. The focus on the effects of colonial domination is even stronger in the neighboring country of Algeria, occupied by France more harshly and for a longer period—from 1830 up until its liberation, after a bloody war, in 1962. But which aspects of colonialism and its terrible effects might a man such as Husayn Ibn 'Abdallah embody? Is he simply a dignitary of Ottoman North Africa that is undergoing collapse as a new colonial world is emerging?

In fact, Husayn Ibn 'Abdallah's life's journey and the disputes over his estate—which lie at the heart of this book—tell a rather different story about this part of the Mediterranean. By following Husayn and the history of his estate after his death, this book argues that historians must transcend any vision of a modern history of the Maghreb perceived solely through the prism of European colonization. This book does not seek to erase the region's colonial past. Rather, it argues for a reinterpretation of the modern period, and of colonialism itself, based on a careful study of what preceded and overlapped with European colonialism in North Africa from the 1880s until the 1920s—namely, the Ottoman provincial culture developed on the southern shores of the Mediterranean for more than three centuries, beginning in the second half of the sixteenth century. The book therefore advocates for an entangled history of the Maghreb written not only by European colonial powers but also by provincial Ottoman people. Such an approach helps cast the history of North Africa and its colonization in broader terms: it reveals significant financial, intellectual, and kinship networks across the Mediterranean that historians have either underestimated or simply ignored. Such an approach also allows us to explore North Africans' many differing conceptions and experiences of time and history during the colonial period.

How Can We Study a Colonized Society?

Since the various North African countries gained their independence between 1952 and 1962, historians have developed diverse methodologies to interpret the region's modern history. Initially, historians sought to explain how and why European nation-states colonized North Africa and other regions.[5] In such a historiography, mainly dominated by French scholars, the history of the Maghreb as a whole was inseparable from its colonization by three European powers: France, which ruled Algeria, Tunisia, and Morocco; Spain, which partially occupied northern and southern parts of Morocco from 1912 to 1956; and Italy, which colonized Libya from 1911 to 1943 (figure 0.2).

This colonial European framework implied two methodological approaches. First, the history of North Africa was conceived of as a continuation of European nations' histories. Second, the privileged, though not necessarily exclusive, use of colonial European administrative sources to write a history of colonial North Africa diminished the attention given to texts and other versions of history written in non-European languages.[6]

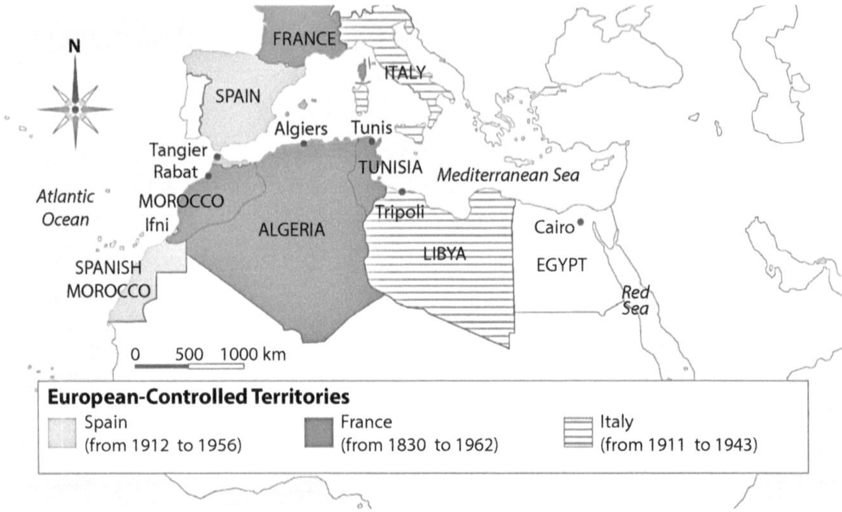

FIGURE 0.2 North Africa under French, Spanish, and Italian colonization

In the case of the Maghreb, privileging colonial sources has not simply led to the disregard of Arabic, the language of Muslim worship, law, and administration in North Africa. Relying principally on European-language resources has also resulted in a neglect of other primary sources written in Ottoman Turkish and Berber, and all the Judeo-Arabic documents (including private correspondence, poems, songs, plays, and accounting records).[7]

At the same time, other historians sought to bring North African societies to the foreground. In the 1960s, Aballah Laroui and Mohammed Chérif Sahli aimed to "de-colonize history" via grand syntheses and historical pamphlets. Point by point, these Maghrebi historians rebuffed what they deemed to be colonialist stances toward North African history.[8] They felt an urgent need to acknowledge the extensive physical and mental violence the colonized subjects had endured for decades, especially in Algeria. Their goal was to deconstruct colonialist discourse that vaunted the superiority of European culture over others.[9]

For their part, anthropologists, sociologists, and other scholars sought to write a bottom-up history of the Maghreb by building on local Islamic and social knowledge. For those who were sensitive to the fact that North African societies had been primarily rural up until the 1970s, it was crucial to explore the world of small towns, villages, tribes, and orality.[10] For the proponents of this bottom-up history, given the dearth of texts produced by the colonized subjects, the most promising methodological approaches were ethnographic surveys. What mattered was to break from a history written solely from the point of view of the state and the urban elites of Tunis, Algiers, and Fez. It was urgent, by that point, to recover the voices of the oppressed peasant masses and understand how they were forced into dependency as the Western capitalist economy made inroads into their territory.

Since the 1990s, both scholarship focusing on colonial domination and studies scrutinizing the social life of North Africans have evolved. Scholars have raised questions about their own methodologies and pointed out some shortcomings of their fields. For instance, historians of European colonialism in North Africa have come to understand how and why the history of colonization has played a central role in the historical approach to modern North Africa and the extent to which this scholarship has focused on Algeria in particular.[11] Three major reasons explain this continued emphasis on colonial history and modern Algerian history. First, more than half a century after the decolonization of North Africa, researchers had and still have at their disposal colonial archives with an enormous quantity of unpublished documents that they can tap into. Second, such a colonial history is still a way for historians to reinterpret and even to challenge the history of the European nations. In the French case, historians of colonial Algeria have brought the French Republic and its much-heralded secularism face-to-face with evidence to the contrary, showing that France did not truly apply its democratic principles and "laïcité" (the French version of secularism) in its colonies.[12] Finally, since the 1970s, studies on the phenomenon of colonization have elucidated the mechanisms of domination present within North African society itself. These works have deconstructed the discourse devised by nationalist leaders to combat the colonial enemy and then to establish their nationalist and authoritarian regimes over the North African populations.

However, even as colonial history came to dominate the field, historians of modern North Africa have realized that this focus on colonization did not come without a cost. The past two decades have seen much discussion of the extent to which the European historical lens has distorted our understanding of the modern history of the Maghreb. The emphasis on French colonial occupation in Algeria, for instance, is so significant that historian James McDougall has acknowledged that the "history of colonial Algeria is, first . . . primarily French history, and then the history of France-and-Algeria."[13] The major actors in this history—namely, the colonized subjects—are not forgotten, but while "they played an active and often decisive role in colonial societies," they are frequently viewed from afar and are not the protagonists of the story.[14] Moreover, as historian Julia Clancy-Smith argues, "the construction of a temporal binary of before and after, precolonial and colonial" has called forth a "dichotomous taxonomy" opposing "Europeans" (colonizers) and "Muslims."[15] Such periodization has led to forgetting other protagonists, located in between the colonizer and the colonized, who did not always take part in these confrontations.[16] Thus, large swathes of the history of early modern North African societies and their interactions with European societies were neglected, despite innovative scholarship conducted in the Maghreb about the precolonial period.[17]

Clancy-Smith's conclusions are in line with the major evolutions in anthropological research on the social life of North Africans. Since the 1980s, work on languages and representations of rural and urban North African populations has gone beyond the idea of homogenous and united masses of colonized peoples. This scholarship has considered the differences between various North African communities, the ties these communities had with collective memory and history, the subsistence of vernacular tongues and cultures (especially Berber-Amazigh ones), and forms of local politicization.[18] Over the course of the 2000s, in the wake of "subaltern studies," scholars again asked questions about the conditions in which colonized populations produced written sources and raised their voices.[19] Isabelle Grangaud, an historian of Ottoman Algeria, suggested that rather than seeking out the voices of the masses, it was more urgent to understand how people were involved in the writing of specific documents making legal claims, and how their claims and struggles could help us reconsider the social history of North Africa.[20] In other words, such local sources do exist, but historians have to pay closer attention to the contexts in which they were conceived and produced.

These same scholars have been instrumental in proposing new paths to explore North African time and space. Some of their works have usefully blurred the distinction between the precolonial and colonial eras to investigate how North African populations adapted to colonial encounters and domination throughout the nineteenth and the first half of the twentieth century.[21] In terms of space, these historians have advocated for a history that would locate the Maghreb in a broader "geography of powers," at the crossroads of various Mediterranean empires and migrations. They have revitalized the history of this region by looking closely at the movements of populations across the Mediterranean, decentering the previously Eurocentric history of migration by arguing that Europeans were not the only ones to travel and discover the world.[22] Historians such as Jocelyne Dakhlia have reestablished symmetries and highlighted reciprocities between the European and Muslim worlds, especially for those Muslims who came to live in early modern Europe.[23] In the same vein, this past decade has seen expanding numbers of studies of the North African Jewish communities, initiating inspiring debates about the many social and political adaptations of these communities in the modern Maghreb.[24]

European Colonization in Light of the Ottoman Legacy

This book aims to explore and expand upon these new methodologies by looking at the history of Tunisia from within and without, before and during the colonial period. While this study centers on the actions and history of exiled Tunisians, it does not only deal with the history of Tunisia. Instead, it relocates the history of exiled Tunisians within North African or Maghrebi societies. The two terms—"Maghreb" and "North Africa"—are arguably fundamental, because they force us to apprehend this area as a single block, without fragmenting it into national pieces.[25] The word "Maghreb," meaning "the lands of the setting sun," was coined during the early medieval period under the Abbasid caliphate. The term includes all the lands located within a so-called *jazīrat al-maghrib*, or "island of the Maghreb" between two seas, the Mediterranean and the Sahara (a sea of sand). It was only during the colonial period that new, European authorities privileged the expression "North Africa" for the French-administered lands of Morocco, Algeria, and Tunisia: *Afrique du Nord* in French, *Africa del Nord* in Italian, and *África del Norte* in Spanish. European colonial authorities did not want to favor the Arabic and Islamic notion of Maghreb promoted by North African nationalists (particularly beginning in the 1920s) to signal the unity of their political cause.[26] Instead, the colonial European administrators developed their own understanding of the space. Still, both notions (Maghreb and North Africa) denominate societies sharing at least two specific characteristics: an attachment to a predominant interpretation of Muslim law (the Maliki interpretation) and the dissemination and intermingling of Berber and Arabic languages.

The Maghreb or North Africa comprises a heuristic unit of analysis, as the four countries forming the region (Morocco, Algeria, Tunisia, and Libya) underwent somewhat similar experiences throughout the second half of the nineteenth and the first half of the twentieth century. Algeria, Tunisia, and Libya formed the western provinces of the Ottoman Empire (*Garp ocakları* in Ottoman Turkish)[27] before being colonized by France (Algeria, Tunisia) and Italy (Libya). The modern history of Morocco differs slightly from that of the three other North African countries, as Morocco was never under Ottoman tutelage. However, this sultanate had a common historical trajectory with Tunisia. In Tunisia in 1881 and in Morocco in 1912, France implemented a similar colonial protectorate regime, whereby France took charge of the military, finances, and diplomacy while the Tunisian and Moroccan sovereigns retained authority over their respective subjects. Given this context, the Tunisian case can and should be analyzed in relation to other North African situations to establish a broader regional history that reveals the interdependence between the "mainland" and "colonies," between "national" and "imperial" histories. More importantly for our argument, the North African peoples—be they from the former provinces of Tunis, Algiers, and Tripoli or from the Moroccan sultanate—had many opportunities to create bonds with each other. Moroccan, Algerian, and Libyan natives made their living in nineteenth-century Tunisia, and vibrant Tunisian communities existed all across North Africa, including in Egypt.[28] Therefore, a history of exiled Tunisians implies the recovery of these networks involving various North Africans across the Mediterranean.

But this book does not limit itself to North Africa. In the wake of the scholarship focusing on the Mediterranean in the nineteenth century,[29] it seeks to foreground the many places in which North African populations acted, in and outside the Maghreb, across the Mediterranean, and throughout the Ottoman Empire. Outside the Maghreb, the book not only takes into account the colonial mainland (here, France) but also shows how exiled Tunisian notables such as Husayn and their entourages were active in Italy, Istanbul—the Ottoman mainland—and Cairo, another major Ottoman city that hosted strong communities of Maghrebi traders and students. North Africa did not differ from other Ottoman regions such as the eastern Mediterranean. It was connected to these other parts of the world, rather than falling under "exceptionalist narratives."[30]

Besides this connected approach, the book aims at understanding the early decades of North African colonization in the light of late provincial Ottoman political culture, following the traces that the Ottoman presence left in North Africa in the second half of the nineteenth century, when this part of the world was gradually coming under European domination.[31] It is unfortunately rare for scholars to consider colonial times in light of what preceded them and, more generally, to think through the history of the Maghreb in terms of the numerous remnants of the Ottoman presence. Despite the Ottoman sultans' wielding tutelage over the region for four centuries, the extent to which the Maghreb was indeed Ottoman is still insufficiently acknowledged (figure 0.3).

Most historical studies still represent the Ottoman provinces of Algiers, Tunis, and Tripoli as autonomous regions, independent of Istanbul and the sultans' power. Autonomy was indeed an ambition of the Ottoman governors, or *beys*, of Tunis in particular between the 1830s and 1870s, and for the *beys* of Tripoli up until 1835, when the Qaramanli Tripolitan dynasty was deposed and the *beys* replaced by governors from the center of the Ottoman Empire. But during the second half of the nineteenth century, European

FIGURE 0.3 North Africa under Ottoman rule around 1830

diplomats (followed by the French and Italian colonial authorities) overestimated this desire for autonomy and established it as an untouchable principle, thereby seeking to preclude and delegitimize any Ottoman intervention in Algerian, Tunisian, and Libyan affairs. The Maghreb nationalist movements corroborated the French view of weak relationships between Istanbul and its provinces. After the 1920s, they gave credence to the idea of the Maghrebi Ottoman provinces' supposed autonomy to support their own claims for independence. This insular vision of the Maghreb during the Ottoman period is still predominant among North Africanists who restrict their field of inquiry to the former provincial realm without properly acknowledging the larger imperial environment. A division of labor has even persisted in European and American universities between Ottomanists, who tend to view the history of the Ottoman Empire through Ottoman Turkish sources, and North Africanists, who still consider the Maghreb far removed from Istanbul and the Muslim world in general.

In reassessing the Ottoman-ness of the Maghreb, this book builds on scholarship begun in the 1970s[32] and takes into account questions raised about the colonial dimension of the late Ottoman Empire in the second half of the nineteenth century.[33] It applies to North Africa a line of inquiry about the lasting effects of the Ottoman legacy that has already led to a reassessment of the history of the Levant.[34] It shows in concrete terms how the effects of having belonged to the Ottoman Empire persisted into the late modern period, during the early days of French colonization in Tunisia. This book argues that North Africa was not just the borderland of an (Ottoman or colonial) empire; rather, it was one of many places for clashes and encounters between competing imperial ambitions. But in order to avoid a distant, top-down view of this Ottoman legacy, this book observes the minutiae of the imperial transition alongside those who experienced it, through a case study: that of the manumitted slave and former Tunisian general Husayn Ibn 'Abdallah and the disputes surrounding his estate.

Husayn: A Transimperial Life and Death

Husayn's case came to light in all its richness and complexity while I was exploring the lives of specific servants to the governors of Ottoman Tunis between the seventeenth and late-nineteenth centuries.[35] These servants,

FIGURE 0.4 Tunisian mamluks during the reform era, Qsar al-Sa‘id Museum (Painting by Alfred Couverchel, 1866–7. Institut National du Patrimoine, Tunisia)

called *mamlūks* (the "owned people" in Arabic), were slaves from Caucasia who had converted to Islam, captives from Italian and Greek households, and freeborn Muslims from the province of Tunis. They had roles as diverse as their origins, from cook to general, gardener to prime minister. The mamluks therefore formed a very flexible political and social category that could help their masters, the Tunisian governors, interact with men from different social classes, from urban notables to peasants.

Following the introduction of Ottoman reforms in Tunisia from the 1830s to the 1860s, the mamluk generals and ministers also played a leading role in adapting the local political culture from patronage and household connections to a more depersonalized administrative service (figure 0.4).[36] With the generalization of public bureaucracy and the end of the slave trade, however, the mamluks vanished. By the 1880s, the last mamluks of the upper echelons of the provincial administration were dying of old age, either in Tunisia where they had served, or else far away in Istanbul, Cairo, or Italy—as in the case of Husayn.

Among this last generation of mamluks, straddling the era of Ottoman reforms and the early days of French colonization in Algeria and Tunisia, the individual whose trajectory struck me as the most intriguing was without a doubt General Husayn Ibn ʻAbdallah (late 1820s–1887), although not because he was an exceptional mamluk. Like many of his peers, Husayn was a slave originating from Circassia (in the northern Caucasus), sold in the markets of Anatolia and taken to Tunisia before being promoted as an administrator to the *beys*. The second part of his name, Ibn ʻAbdallah (son of the servant of God), was often applied to converts to Islam in order to erase any reference to their non-Muslim fathers' or mothers' names.

What drew my attention, rather, were the circumstances surrounding his death and its aftermath. Unlike other mamluk dignitaries from Tunis, Husayn did not die in Tunisia or any other part of the empire still under Ottoman tutelage, such as Cairo or Istanbul. When Husayn passed away in 1887, six years after the establishment of the French protectorate over Tunisia, he was living in Florence, having represented the interests of the Tunisian state in various financial disputes in Tuscany since the 1870s. While we know very little about the estates of many Tunisian dignitaries, there are exhaustive primary sources about the disputes that sprang up around General Husayn's estate in Tuscany from 1887 to the early 1890s. The primary sources from the French, Italian, Tunisian, Ottoman, and English archives are particularly rich and varied, both because of the administrative and financial interest of the papers drawn up by Husayn and because of the intensity of the competition between the French, Italian, and Ottoman authorities over Tunisia during this period.[37]

One could argue that this example is too modest and too specific to the dying world of Ottoman dignitaries. From such a perspective, this apparently modest case could not contribute much to a reinterpretation of modern Maghreb history. However, a case must be judged not solely "on the basis of the size of the object examined," but on what it can reveal from a historical and methodological viewpoint, and on the reverberations it once sent across societies.[38] Thus, the circumstances surrounding Husayn's legacy are significant in that they offer a "research tactic" and a "new framework for reasoning."[39] To a certain extent, Husayn's case can be compared to the "imperial" and global lives of other historical characters spanning worlds and languages, revealing, via series of connections, unexpected dimensions to the history of their times.[40] Husayn's life, afterlife, and especially the story of his estate cast light on a specific, though very important, aspect of this global history: it reveals the extent to which Tunisian, North African, French, Italian, and Ottoman histories—histories that are often studied separately—were in fact deeply entangled.[41]

More significantly, following the trajectories of "socially situated individuals" forces us to stop thinking of the territories, categories, and languages within which people move as something predetermined.[42] Thus, while situating the action between Ottoman and European lands, I do not presume any particular continuities or discontinuities between these political entities.[43] I do not believe these worlds to be antagonistic, irreconcilable, or distinct in nature. What I hope to achieve by making intensive use of Husayn's case study is to reconstitute social worlds and networks that tend to be hidden from view when colonization is the sole focus. The conflicts over Husayn's estate were not limited to a small circle of people. They involved lawyers, Italian senators, Parisian bankers, Husayn's own Tunisian servants in charge of keeping the accounts of his properties in Tunis, and Jewish, Muslim, and Christian tenants of Husayn's properties near the Tunisian port of La Goulette (*Halq al-Wadi* in Arabic), as well as members of Sufi brotherhoods dotted around Morocco, Egypt, and Istanbul. By recovering the connections between these social universes through Husayn's case study, my purpose is not only to present various interpretations of the modern history of the Maghreb, or to call into question some grand narrative entirely ordered around the ordeal of colonization—I also hope to set out new lines of historical interpretation.[44]

This study differs from other works about imperial and global lives, as my primary purpose is not to reconstitute Husayn's life.[45] Husayn's existence will constitute the core of only the first chapters of this book. In the second half, I examine his death—and, more specifically, the disputes and claims surrounding his estate—in order to reconsider our historical interpretations of this part of the Mediterranean.

The first chapter explores Husayn's life from his arrival in Tunis as a slave in the 1820s to his exile in Tuscany in the 1880s. This chapter departs from prevalent interpretations of Tunisian and modern North African history more broadly, which stress either colonial domination or a national framework of analysis. I build on recent innovative scholarship employing a Mediterranean or internationalist perspective to show that neither the main stages in Husayn's life nor his choices can be interpreted exclusively in terms of the increasing confrontation with Europe or in terms of nascent state-building. The stages of Husayn's life take on their fullest meaning in light of questions that arose in all empires. For instance, the abolition of slavery did not depend solely on European intervention in the Muslim world. Through Husayn's experience of slavery, one can understand that the issue of abolition involved local actors and was strongly shaped by debates about the forms of princely service, especially in relation to the rise of a public bureaucracy affecting other administrations around the world.

The second chapter looks in greater detail at another dimension of Husayn's journey and experiences—the highly localized yet transimperial nature of debates about private and public property—by examining Husayn's assets, comprising real estate, movable goods, and credit. Historians have correctly perceived Maghreb subjects as having been dispossessed and reduced to material poverty in the colonial context.[46] But this history of dispossession should not blind us to other transimperial transformations that had been set in motion at an earlier point during the nineteenth century. Examination of Husayn's landholdings north of Tunis reveals similar issues to those already identified in Egypt, the center of the Ottoman Empire, and tsarist Russia—all places where the ideas of public land and private property were hotly debated by the second half of the nineteenth century.[47] A study of Husayn's assets shows that the history of property and appropriation in the modern Maghreb is not solely a tale of colonial seizure. Husayn and other Tunisian officials were able to take advantage of the various statutes and laws incumbent on landed property and its public and private components. They sought to control European access to their property and to hide some of their assets behind false accounts. Our understanding of colonial dispossession in the late nineteenth century therefore needs to be supplemented by an account of the continual alterations to land ownership rights and the increasing reconfigurations of public land and private property that preceded and continued throughout the colonial and postcolonial eras on a global scale.[48]

The third chapter expands this perspective as it follows a later stage of Husayn's career in the wake of two major affairs relating to the assets of two other Tunisian dignitaries—Mahmud Ibn 'Ayyad and Nessim Scemama, former directors of Tunisian finances. Both fled to Europe after embezzling Tunisian public funds, one in the mid-1850s, the other in the mid-1860s; Husayn, representing the Tunisian state, mediated the resulting financial disputes, first in Paris and then in Tuscany. Husayn's case, as well as these two major financial scandals, exemplifies a crucial yet understudied phenomenon of the nineteenth century: the increasing legal and financial engagement between Europe and Tunisia before its colonization. Moreover, in such cases, the increasing use of the notion of "affairs" to refer to such cases (*nazīla* or *qaḍiya* in Arabic, *affaire* in French, and *affare* in Italian) points to the spread of this moral category across the Mediterranean, a dynamic that scholarly literature on the Mediterranean has not previously acknowledged.[49] While Europeans used the term "affair" to criticize Tunisian mismanagement and so-called North African corruption, Ottoman dignitaries such as Husayn used the same notion as a political resource to fight their European rivals. Lastly, this third chapter shows that disputes over the estates of Tunisian dignitaries contributed to the great increase of written evidence from both sides of the Mediterranean in the era of print, contradicting the idea of a dearth of North African sources often encountered in studies about colonial North Africa.

The second part of the book investigates the disputes surrounding Husayn's estate and their outcomes after Husayn passed away, in order to reveal how colonized North African Muslim and Jewish individuals kept or reshaped their agency beyond the colonial scene and across the Mediterranean. Chapter 4 focuses on diplomatic conflicts over Husayn's estate in 1887–88 involving the Ottoman Empire, on the one side, and the French and Italian authorities, on the other. Disputes between the French and Ottoman consuls over Husayn's papers, bodily remains, nationality, and especially his burial in Istanbul during the summer of 1887 belie the stereotype of an Ottoman Empire too weak and remote from the Maghreb for its sultans to accord it much interest during the late nineteenth century.[50] The same chapter explores Italian authorities' intervention in Husayn's affairs in late 1887, in which they sent judicial officers and police to the chancery of the French consulate in late 1887 to seize papers relating to Husayn's estate after a creditor (who was also Husayn's former right-hand man) filed a complaint. Italian official and legal involvement in Husayn's affairs acts as a reminder of the extent to which subjects from Algeria and Tunis used Tuscany, and Italy more generally, to conduct legal action and appeals. The purpose of this chapter is to reveal a new line of inquiry into a history of North Africa that would take into account the many places used by various Muslim, Jewish, and Christian populations across the Mediterranean to assert their legal and political rights.[51]

The fifth chapter looks at more personal levels of litigation than these diplomatic conflicts. This chapter turns to four groups of individuals who claimed portions of Husayn's estate: the former *bey*, or governor, of the province of Tunis, who argued for his family's rights of ownership over the emancipated slave Husayn; Angiolina Bertucci and Eva Keusch, the Italian and Franco-German mothers of two girls that Husayn raised in the Tuscan city of Livorno; a well-known Tunisian scholar (Shaykh Bu Hajib) and his sons, who were close to the deceased; and an Algerian Jew, Léon Elmilik, who served Husayn for a decade. Studying each of these groups allows us to engage with major fields of modern North African history—namely, the history of Muslim elites, of women, and of North African Jewish communities. This chapter demonstrates that the *beys* of Tunis were not impotent sovereigns placed under French tutelage, as is widely believed. They continued to govern informally from their palaces, drawing on trusted mediators in Istanbul and Italy. It further shows that European women did not defend their rights using only colonial institutions; it was precisely by presenting herself as part of the Muslim line of descent of the Bu Hajib family that Italian mother Angiolina Bertucci managed to put forward her claims more effectively than Franco-German Eva Keusch. In the third section of this chapter, the disputes between the Muslim scholar Bu Hajib's family and the *beys* over the properties in La Goulette underscore how vigorous debates in Arabic could transform Tunisian justice, a fact that colonial historiography about the Maghreb rarely takes into account. Lastly, Elmilik, with his multiple proceedings and claims, is an example of those Maghreb Jews who cannot be studied exclusively within the framework of their community nor solely as assimilated French citizens. They instead need to be understood in the context of belonging to multiple groups and many places, as recommended by historian Daniel Schroeter.[52]

As an epilogue to this story, the final chapter reveals what became of Husayn's estate. Which of the legal heirs and claimants won? And to what extent did this affair contribute to transforming the colonial system, legal practices, and archiving processes during the colonial period? I track what became of the successful claimants, Shaykh Bu Hajib's household, from 1890 to the 1920s to illuminate social and political dynamics that historians of modern North Africa have underappreciated until now. The Bu Hajibs' control over Husayn's real estate at La Goulette and the integration of the Italian mother Angiolina Bertucci and her daughter into their household boosted the Bu Hajibs' capacity for action and helped them extend their reach on various fronts in the Muslim, Ottoman, and colonial worlds. From the 1890s to the 1910s, Shaykh Bu Hajib was deeply engaged in Islamic reform and debates with other Muslim scholars—matters that did not always involve colonial authorities. On the other hand, at the beginning of the twentieth century, the Italian girl raised by Husayn became a member of the shaykh's household. As such, she strengthened the Bu Hajibs' ties with young Tunisian reformists and with the Ottoman Empire in its last days. She went to Istanbul in 1912, in the wake of the Young Turk Revolution, where she joined her husband, Tunisian reformer 'Ali Bash Hanba, who was head of the Special Organization (*Teşkilat-ı Mahsusa*) involved in the Armenian genocide and Muslim uprisings against the Western Allies in Istanbul during the First World War. Lastly, competition among the Tunisian *bey*'s entourage lifted one of the Bu Hajib sons to the post of prime minister of Tunisia from 1926 to 1932. Previous historical research did not take these competitions for power seriously, with the Tunisian sovereigns' entourage being viewed simply as an elite group of "collaborators" in the service of colonial authorities. My detailed examination of the Bu Hajibs' lives thus opens up new avenues of inquiry about the elites involved in the administration of Tunisia during the colonial period.

While the first three chapters of the book resituate the Maghreb in transimperial debates and broader historical dynamics, the last three chapters uncover spaces and forms of action that were crucial for exiled North Africans such as Husayn.[53] Yet the entire book is concerned with forms of writing and archiving in the second half of the nineteenth century. A great diversity of texts, including private and administrative correspondence, legal deeds, telegrams, and travelogues, inform us about Husayn's estate. Records about the "Husayn estate affair" abound in details of how Husayn and his entourage's writings were translated from Arabic into French, summarized, inventoried, and then shipped in trunks to France and Tunis. Some of these documents were then consigned to the diplomatic and colonial archives; others were published as manuscripts and printed editions in Arabic. This shows that the colonial archives, despite being the basis for studies of the modern Maghreb, were not only manipulated and rearranged for political ends—they blind us to the sheer profusion of writing in nineteenth-century North African societies. Husayn's case makes it clear that these societies—just like Europe at the time—produced a prodigious volume of documents in various languages.[54] The wealth of written and printed documents in the Maghreb is another strong indication that we should not restrict our exploration of the history of North Africa to the "colonial moment."[55]

The main argument of this book—namely, that we need to contextualize the history of these colonized societies of North Africa with regard to their Ottoman legacy—does not arise solely from a desire to balance our understanding by adopting a more exhaustive approach toward the modern history of North Africa. The need to bring the Maghrebi people back to the center of North African history is a political necessity. Concentrating exclusively on the colonial period as the prime moment of access to modernity and of full interaction with European societies can only reinforce the now widespread preconception that has taken root in Maghrebi societies: that the Frenchified and Europeanized components within North African societies—which use European languages and adopt customs perceived as being European—stem directly from colonization and are thus external to some supposed preexisting Islamic substrate.[56] Broadening the historical framework and viewing the modernity of Maghrebi and European societies as a conjoined phenomenon ought to dissuade us from segmenting populations on the basis of language or lifestyle. The so-called Frenchified and Europeanized North Africans are no more alien to North Africa, no less a part of its societies, than those who claim that a "genuine" Islamic identity is inherent to the Maghreb. To understand the debates and divergences within Maghrebi societies, we must bring the colonial and so-called precolonial periods together into a single framework.

Moreover, this book's attempt to rewrite the history of the Maghreb—as opposed to the history of European colonization of the Maghreb—offers a way of countering the growing marginalization and dehumanization of Maghrebi immigrants, workers, and their descendants in France and other European countries (including Belgium, the Netherlands, Spain, and Italy) where communities of migrants from the Maghreb have settled over recent decades. Their presence in Europe is not solely the result of colonization but also stems from the earlier relationships between Maghrebi societies and the northern shores of the Mediterranean in the early modern period.[57]

Lastly, by historicizing the specific dynamics of these societies, we bring the Maghrebi populations' ability to act back into view. We had so completely forgotten North Africans' capacity for action and reaction that the Arab Spring came as a complete surprise. We had come to think of these countries as “noiseless,” as too apathetic and exhausted to be able to overthrow authoritarian regimes.[58] But the 2011 revolutions, the 2019 Algerian protests, and the consistently unpredictable ways in which they have played out make it even more important to explore the history of Maghrebi societies, not just during these recent revolutions or the era of European occupation of North Africa, but also prior to colonization.

ONE

Husayn

An Ottoman Reformer and a Product of Ottoman Reforms

> When we reached (the small village) of Qal'at al-Andalus,[1] we found the river flooding. We then had to cross the river on horses. Halfway I could no longer move forward. I had to swim. After having crossed the river . . . I remained in wet clothes for hours, to the point that I felt a pain in my kidney. A pain that I would suffer from over the course of my life.[2]

HUSAYN WROTE these lines in 1886, one year before he passed away. In this letter, sent to a friend and former colleague, the aged mamluk recalled a specific day in the early 1840s, when he was still young and new to the Ottoman province of Tunis. More than four decades after having been brought as a slave from the Caucasus, he still remembered his physical pain. He was full of memories. Husayn might also have remembered other crucial events in his life: his separation from his parents in the late 1820s when he was still a child; his student days a decade later in the first state school in Tunisia; his promotion to general and then to mayor of Tunis in the 1850s; his first voyages to Paris and London; and, beginning in the 1870s, his time in Florence, where he remained after the French takeover of Ottoman Tunisia.

Husayn, like any other human being, carried his various historical experiences with him throughout his life. This first chapter aims to recover those experiences at each major step in his life, from his arrival in Tunis as a slave in the 1820s to his exile in Tuscany in the 1880s. European expansion across the Mediterranean Islamic rim and the French colonization of North Africa from the 1830s onward undoubtedly affected Husayn's trajectory. After the French conquest of Tunisia in 1881, he lost all his official positions and political influence. Indeed, French authorities barred him from returning to Tunis. However, French colonialism was not the only historical element explaining or constraining the actions Husayn took in the course of his life. In this chapter, I argue that many other historical elements help explain his actions, though these same elements often constrained them. Chief among these was the Ottoman Empire's reform program throughout the nineteenth century, as it profoundly shaped Husayn's understanding of history, his political culture, and his sense of belonging. Husayn was deeply involved in reforming Ottoman Tunisia; at the same time, he was the product of the political culture shaped by these reforms. The purpose of this chapter is not to reveal a reality of which Husayn was unaware; rather, by recovering his subjective historical experiences as best we can, it seeks to understand the choices he made and the conditions governing those choices.[3] Understanding Husayn's conceptions and experiences—including his extensive involvement in the restructuring of Ottoman rule at the provincial level—is crucial, because they colored his exile in Italy during the 1880s and his legacy in colonial Tunisia.

In this first chapter, I will examine the two main stages of Husayn's life: his time as a slave and Ottoman dignitary in Tunis and his exile in Tuscany from the mid-1870s to 1887. In the first part of the chapter, I will show how the major episodes of the first stage of his life—from his childhood in the 1820s to his becoming an important statesman in Ottoman Tunisia in the 1850s—are entwined with the implementation of what historians have called the Ottoman reforms (*iṣlāḥ*, *tanẓīmāt*). By the end of the eighteenth century, Ottoman sultans had started to significantly reform their army, administration, and justice system, a reform project whose pace increased from the 1820s onward. The sultans implemented these reforms to strengthen their domains and their state apparatus, which was under attack by the Russians, the Habsburgs, and the French.[4] Yet similar reforms were only enacted in Tunis and the neighboring province of Tripoli beginning in the early 1830s, and the sultanate of Morocco only followed in the 1840s. North African rulers, then, were not simply imitating Ottoman sultans. Their internal reforms were a response to the traumatizing French conquest of Algiers in 1830. From Algiers, the French army slowly expanded its control, colonizing the Algerian coast and threatening to destabilize neighboring Moroccan and

Tunisian territories. Later still, under pressure from France, Great Britain, and Italy (after its unification in 1860), North African rulers implemented major reforms that suited specific European political agendas. For instance, the governors of Ottoman Tunisia abolished slavery in 1846 and enacted a constitution in 1861, mainly under British diplomatic pressure.

I will then turn to the second stage of Husayn's life, beginning in the mid-1870s, when he left Tunis to represent his government's interests on missions in Europe, first in Paris in the 1850s and then in Italy in the 1870s. As the years passed, Husayn spent more and more time in Tuscany, finally going into exile there after the French conquered Tunis in 1881. He was not the only dignitary to leave colonized Tunisia. Four of his closest Tunisian friends and allies, renowned statesmen and Muslim scholars, went into exile in the Ottoman capital, Istanbul, and in other major Ottoman cities such as Cairo and Alexandria. Their shared experiences remind us that here, as in any other context, exile provoked a crucial experience of self-redefinition.[5] It was in exile, as we will see, that Husayn and his interlocutors redefined their own understanding of the Islamic world. And it was in Tuscany that Husayn distanced himself from European Christians and North African Jews. Eventually, he established an atypical lineage: he raised, educated, and eventually passed on some of his estate to two young girls, one of German matrilineal descent, the other born to an Italian mother. The latter would become instrumental in allowing Husayn to exert agency even after his death; it was she who passed on his Ottoman legacy and some of his real estate from the 1890s to the 1920s.

One of the Last Male Circassian Slaves in Tunis

Before he became General Husayn Ibn ʿAbdallah, before being trained and promoted in Tunis, Husayn had lived a very different life. He was born in the late 1820s, very far from Tunis, on the outskirts of the Ottoman Empire, in Circassia, a region located in the northern Caucasus along the Black Sea, between Russia and Georgia (figure 1.1).

Like other Circassians taken as slaves since the sixteenth century, Husayn was purchased to staff the Ottoman palaces in Istanbul and in the provincial cities across the empire. As was often the case in Circassian society, Husayn was either forcefully separated from his family as a child or sold

FIGURE 1.1 Circassia by the mid-nineteenth century

by his own parents.[6] It is possible that, throughout his life, Husayn used a Circassian language when speaking to other slaves hailing from the same region. But as far as we know, he never explicitly mentioned or recalled at length his native culture, his parents, or his homeland. His personal letters from the mid-1850s on are silent on these topics. Like most uprooted Circassian slaves before him, Husayn settled within the Ottoman Empire and was raised to understand this world as his new homeland.

Husayn was one of the last male Circassian slaves to be sold to the Tunisian governors. Ottoman and North African dignitaries continued purchasing female Circassian slaves for use as concubines until the beginning of the twentieth century. However, these same Tunisian dignitaries ceased to bring male Caucasian slaves from Istanbul by the 1830s. The total number of these male slaves, or mamluks, serving the Ottoman governors of Tunis reached its apex in 1826, when they numbered approximately 270, including Circassians, Georgians, Italians, and Greeks. After the 1830s, their numbers slowly dwindled.[7]

There are many reasons for the steady decrease in the mamluk corps, including changing attitudes toward slavery. Western European captives converted to Islam—termed "renegades" in European literature—had comprised a significant portion of the mamluks in North Africa since the medieval period. But in 1816, under European pressure and after a British military campaign in the Mediterranean, North African rulers (including the Tunisian governors) were no longer allowed to enslave Italian, Spanish, or other European prisoners of war. They even had to free their Christian slaves. Two decades later, in the 1840s, British diplomats—who were hugely involved in campaigning for the abolition of slavery across the world—advocated even more actively for the end of slavery within the Mediterranean and the Ottoman Empire. In Tunis, the British consuls convinced the Ottoman governor Ahmad Basha Bey (1837–55; figure 1.2) to shut down the slave markets in 1841 and, five years later, to manumit slaves and abolish slavery entirely.[8] At that time, however, the abolition of the slave trade mainly targeted African slaves in Tunisia and across the Ottoman Empire.

FIGURE 1.2 Ahmad Basha Bey, Tunisian ruler (1837–55), Qsar al-Sa'id (Institut National du Patrimoine, Tunisia)

The gradual restriction of the Circassian slave trade was actually the outcome of changes within the Ottoman Empire. The trade in Circassian slaves underwent profound changes and eventually dwindled in the two decades following 1864, when Russia took over Husayn's homeland. Because of the Russian invasion, the Ottoman government had to deal with a flow of Circassian refugees and the slave trade they brought with them. This trade endangered public order: historian Ehud Toledano has referred to "violent clashes" that erupted in Anatolia "between Circassian slaveholders and their slaves." These clashes were so violent that they prompted Ottoman authorities to abolish the Circassian slave trade within their borders.[9]

In Tunisia specifically, the slow death of the trade in slaves from Circassia and the Caucasus more broadly was also related to local changes. The Ottoman governors of Tunisia, for instance, no longer needed to rely on mamluks for administrative and military tasks. As in other parts of the Ottoman Empire, by 1831 the Tunisian governors had created a new, regular army, also called *niẓām al-jadīd* (the "new order"). Far from being only slaves, these new soldiers were free men conscripted to serve the Tunisian state. Similar changes occurred with the civil administration. Tunisian governors no longer preferred personal, committed servants such as the mamluks. By the end of the 1850s, the Tunisian state had created new civil institutions, including ministries and central and local civil courts, and hired civil servants working under a new, formal hierarchy to staff them.[10]

Even mamluks like Husayn started to redefine themselves not as personal servants or slaves of the Ottoman governors of Tunis, but as civil servants. Husayn had no difficulty labeling himself a "mamluk,"[11] nor did he, as a member of this group, cease acting for the "interest of his government" and the "love of his country."[12] Yet Husayn expressed his explicit support for the abolition of slavery on a global scale in a letter he wrote in 1864 in response to a query from the U.S. consul in Tunis, Amos Perry. As the Civil War raged in the United States, Consul Perry wondered about "the effect of slavery" on Tunisia: "whether its abolition was met by Tunisians with sorrow or with joy" and "which of the two kinds of service has been proved by experience to be more suitable, compulsory service, i.e., the service of slaves without pay, or voluntary service, i.e., service for specific pay."[13] In his reply, later printed in Arabic and in French "in some half dozen different papers" including the journal *L'Europe* in July 1865, Husayn did not simply explain how Tunisia had ended its slave trade.[14] He argued that humans should outlaw slavery for economic reasons, as a matter of efficiency. Building on the rationale provided by "specialists in political economy" (*al-iqtiṣād al-siyāsī*) and following, in a way, Thomas Jefferson's idea that slaves tend to decrease their master's industriousness,[15] Husayn noted that "nowadays, countries where full liberty exists and no enslavement is permitted are more prosperous than other countries":

> Those who used to depend on slaves for service and failed to hire free men returned to the natural way and desirable behavior, for they realized that man should meet his necessary needs by himself and decrease dependence on his brethren because if one becomes used to hiring others, one might end up incapable of fulfilling the least of necessities. . . . I myself believe that universal liberty and the non-existence of slavery have a deep effect on refining a man's manners as well as on the development of culture.[16]

The abolition of slavery in Tunisia, then, was not due solely to British pressure or European imperial agendas. The mamluk corps disappeared with the ending of the Caucasian slave trade, as civil servants slowly replaced personal servants and slaves within the Ottoman administration and armies.[17] Husayn might have been aware of the British abolitionist campaigns within the Ottoman Empire; he certainly knew about abolitionism and the Civil War in the United States, as his correspondence with the U.S. consul attests. But as a slave and servant, he personally experienced a profound, internal reconceptualization of the meaning and practice of slavery and state service during the reform era. This reshaped his sense of self and his agency, both before and during the colonial period.

One of the First European-Educated Slaves in Tunis

Husayn was a product of the Ottoman reform era in yet another sense. Soon after he arrived in Tunis, Husayn, like all the other mamluks serving in the Tunisian governors' palaces since the seventeenth century, studied the Quran in order to learn to read and write in Arabic, one of two main administrative languages in Tunis, Tripoli, and Algiers—the other being Ottoman Turkish.[18] But unlike older mamluks and even some mamluks from his

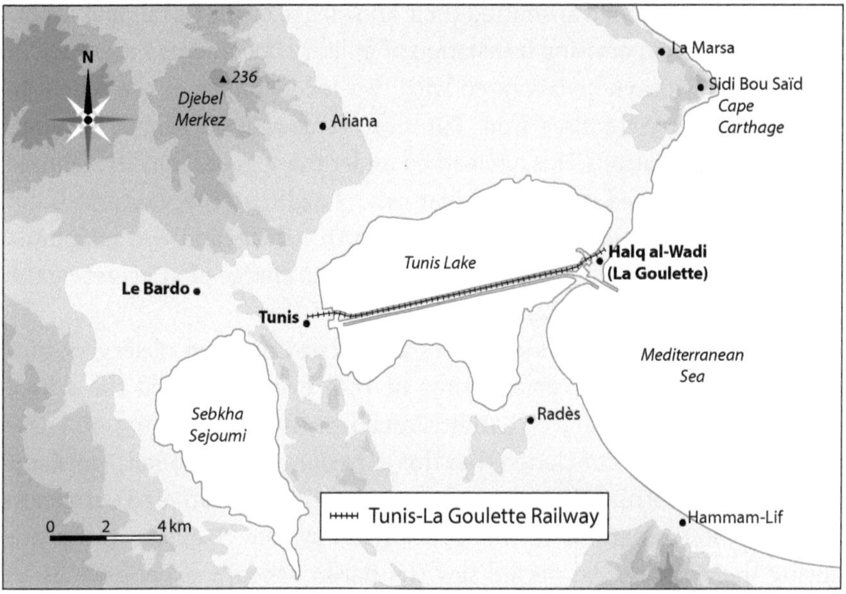

FIGURE 1.3 Tunis and its outskirts

own generation, Husayn went further than simply studying the holy book under the supervision of an instructor (*mu'addib*). He was one of the very few mamluks to attend a new educational institution, the Tunisian Military School (*maktab al-'ulūm al-ḥarbiyya*), alongside free Tunisians. Tunisian governor Ahmad Bey had founded the Bardo Military School, located within the governor's palace in Bardo, in the western outskirts of Tunis (figure 1.3), in the late 1830s, six years after the implementation of similar educational establishments in Istanbul.[19]

Like these other Ottoman institutions, the Bardo Military School was devised as a place to train young men to become officers and civil servants in charge of the new army, administration, and state courts. The Bardo Military School also followed the example of the French École Polytechnique, a military academy established in Paris in 1794.

Until the late 1860s, European military officers at the Bardo School taught French, mathematics, and the new arts of war, ranging from military drills and engineering to artillery and the art of fortification. Islamic scholars (*'ulamā'*) at the school transmitted their knowledge of the Quran and Arabic grammar while supervising translation of military documents into Arabic.[20] In the 1840s, Husayn collaborated with two fellow pupils at the school to translate military treatises from European languages to Arabic. To complete these translations, Husayn worked under the aegis of Shaykh Mahmud Qabadu, a renowned Islamic scholar who taught at the Zaytuna Great Mosque—a major Islamic institution in North Africa—and was appointed judge (*qāḍī*) at the Bardo palace and then *muftī* (in charge of interpreting the law) from 1867 until his death four years later.[21]

The school's European instructors as well as European observers considered it a failure. A French visitor in 1844 even compared the school to "a den of debauchery."[22] Earlier French historians, building on these European accounts, concluded that this experience exemplified Tunisians' inability to modernize their country.[23] As a counter-narrative to European-centered interpretations of this educational system, Tunisian historians during the postcolonial period saw the Bardo School's experience as an early embodiment of the Tunisian nation. According to this nationalist interpretation, it was the Bardo School that, for the very first time, brought together Tunisians of various backgrounds—from the city of Tunis and the countryside, slaves and free men—to govern their country's new institutions.[24] In any case, Husayn's education at the Bardo School demonstrates that the school should not be understood as a failed attempt at modernization or a significant episode in the process of nation-building, because the military officers and administrators trained there still held their positions during the first several decades of the French protectorate in Tunisia.[25] Nor did the Bardo School revolutionize the transmission of knowledge, miraculously sowing the seeds for the formation of the Tunisian nation. Instead, it seems closer to the truth to simply say that the school allowed its students to combine Arabic and Islamic knowledge with European administrative expertise.

Armed with this education, unlike his mamluk predecessors, Husayn knew more than simply reading and writing.[26] He had acquired a command of old and new narrative techniques. To express his political ideas, he used older local genres such as the open letter (*risāla*). Four years before his *risāla* on slavery to the U.S. consul in 1864, Husayn had written another major letter, this time on the political rights of Tunisian Jews.

In this letter, he stated that Tunisian Jews did not necessarily have to be represented on the new Supreme Court that monitored the Tunisian budget and lower courts from 1860 to 1864. Its members were selected on their merits, not their origins, he argued, and Tunisian Jews had to learn to prove their loyalty to the Tunisian state.[27] He also tried out new European narrative forms for the Arabic-speaking elites in the province of Tunis. He produced what is viewed as the first children's tale set "in modern Tunisia," in which the characters are meant to resemble contemporary people.[28]

Like previous mamluks, Husayn relied on scribes born in the province of Tunis who had studied at the Zaytuna Great Mosque to carry out his administrative duties and write his letters.[29] But thanks to his education at the Bardo School, Husayn was far more autonomous than his predecessors. With his ability to write in a variety of styles and genres, Husayn blurred the old division between mamluks, who gave orders, and scribes, who wrote them down. Until the end of his life, when not afflicted by rheumatism, Husayn would compose and sign his own letters in Arabic without systematically relying on his scribes (figure 1.4).[30]

Building on his knowledge and extensive literacy, Husayn acquired an authoritative position among the scholars and authors of his time. Like older mamluks, who prided themselves on their love of books, Husayn was at the center of many exchanges and loans of written works.[31] He was a patron to *'ulamā'*, and poems were written in his honor.[32] Yet he surpassed his predecessors in some respects, as his voracious appetite for writing and debate placed him within a new circle of officers and Muslim scholars. Thanks to his close relationships with Shaykh Qabadu from the Bardo School and another mamluk of Circassian origin, Khayr al-Din, who became his patron

FIGURE 1.4 Husayn's signature combining the letter "H" from the Latin alphabet and the Arabic *ḥa* followed by a "y."

FIGURE 1.5 Khayr al-Din, Qsar al-Sa'id Museum (Painting by Louis Simil, 1852, Institut National du Patrimoine, Tunisia)

(figure 1.5), Husayn became part of a network of intellectuals that included three scholars who had also studied under Qabadu and gone on to become leading thinkers in Tunis in the second half of the nineteenth century.

The first scholar, Muhammad Bayram V, came from a renowned family of Islamic jurisconsults. He adhered to the Hanafi legal school that was influential among Ottoman elites in North Africa.[33] The second, Salim Bu Hajib, Husayn's Arabic teacher, taught at the Zaytuna Great Mosque in Tunis from 1848 to 1917. Husayn employed Bu Hajib as his secretary from 1858 onward.[34] The third, Muhammad al-Sanusi, born in 1851 to a family of *qāḍī*, was one of Bu Hajib's many students; he acted as secretary (*kātib*) and then as editor of the state newspaper *Rāʾid al-Tūnisī*, founded in 1860.[35] Moreover, Bayram V, Husayn, and Bu Hajib assisted Khayr al-Din in writing a political treatise in Arabic titled "The Surest Path to Knowledge Concerning the Condition of Countries."[36] The Bardo experience thus allowed officials like Husayn to redefine their relationship with Islamic scholars and therefore to reshape the crucial relationship between power and knowledge (*sulta* and *ʿilm*). By exposing pupils to the combination of old and new narrative techniques, the Bardo School also shaped the way that men like Husayn would reframe their understanding of social realities.

Advocating Over and Over for Reforms

Husayn's role within this intellectual circle of statesmen—in addition to his Bardo and Islamic training—hastened his political ascent within the new administrative institutions in Tunis following a second wave of state reforms at the end of the 1850s. Under the protection of Khayr al-Din, Husayn became the first chairman of the municipality of Tunis, created in 1858.[37] Two years later, he was appointed the head of the new department of foreign affairs and, in 1861, became the chairman of a new civil and criminal court in Tunis that was part of a new constitutional system enacted the same year.[38] Not only did Husayn push for legal and administrative transformations while serving in these posts, he was also instrumental in literally spreading the news. In 1859, along with British trader Richard Holt, Husayn played a major role in setting up the first Tunisian state-sponsored newspaper in Arabic, *al-Rāʾid al-rasmī*, also titled *al-Rāʾid al-Tūnisī*. Husayn funded the importation of type blocks from Europe. The newspaper had a print run of 1,500 copies, half of which were for Tunisian bureaucrats who were required to take out a subscription.[39] The first half of the paper covered new reforms and laws enacted by the state; the second was devoted to international news.

By the beginning of the 1860s, Husayn was so adamant about the need to implement reforms in the Ottoman Empire that he felt a kinship with other military and modernizing empires, including the Second French Empire, which was by then reforming the French colonial administration

FIGURE 1.6 Napoleon III meeting Muhammad al-Sadiq Basha Bey in Algiers in 1860, Qsar al-Sa'id Museum (painting by Alexandre Rebelle, 1862, Institut National du Patrimoine, Tunisia)

in Algeria. In a written report of Tunisian governor Muhammad al-Sadiq's visit to Algiers to show a draft of the Tunisian constitution to Napoleon III, Husayn portrayed himself as more culturally akin to the French emperor and Empress Eugénie than to his fellow North Africans (figure 1.6). For instance, when recounting a reception hosted by Arab tribesmen, he took pains to distance himself from the colonized Algerians and their manners, which he depicted as unsophisticated:

> Behind the tent, a reception hosted by the Bedouins was taking place in honor of the king and the queen. It was full of dishes of couscous and meat as well as roasted mutton that had been laid out on tree trunks and pieces of wood, with their intestines still in their bellies. Someone with a voracious appetite would have left nothing of the mutton but its sweaty skins, which had been cooked in an out-of-date and savage way. . . . The queen [Empress Eugénie] felt surprised

seeing this strange scene, she began to smile. . .]. She said to our master [Muhammad al-Sādiq]: "Are you enjoying this strange sight?"[40]

Less than four years after this expression of sympathy with the French emperor and empress, however, Husayn's efforts to reform Tunisian politics were undermined. The Tunisian government had to fiercely suppress a province-wide revolt. In 1864, rebels demanded an end to legal and tax reforms that they neither understood nor considered consistent with their Islamic values.[41] The Tunisian governor and his prime minister, the Greek mamluk Mustafa Khaznadar, revoked the 1861 Tunisian constitution and new institutions that followed it.[42] Mustafa Khaznadar sidelined Husayn, his patron Khayr al-Din, and their close allies, who were advocating for the reforms and who criticized the harshness of the state's repression of the uprising. Their roles in the Tunisian government thus marginalized, Husayn and Khayr al-Din resumed their travels in Europe and elsewhere.[43]

After going to Istanbul, Husayn visited Germany, Sweden, Denmark, Belgium, France, Italy, England, and Spain. He even traveled to New York and Washington in the summer and fall of 1867.[44] Before crossing the Atlantic Ocean, Husayn studied the U.S. institutions and asked for a recommendation letter from Amos Perry, the American consul in Tunis, who praised Husayn as the "most enlightened *musulman* person in Tunis."[45] According to Perry, "a young man of decided literacy, taste, and historical attainments" named Nonce Rocca was supposed to accompany Husayn.[46] However in August 1867, on the ship *Scotia*, it seems that Husayn was the only passenger sailing from Liverpool hailing from Tunis-"Turkey" (meaning the Ottoman Empire). On the passenger list, his name appears as "Hassein Huessin"—a repetition and variation of his first name without any reference at all to his other name, Ibn 'Abdallah (son of the servant of God), often used in Muslim societies to replace the memory of converts' and slaves' non-Muslim parents.[47]

Husayn visited the United States as a private citizen and, to our knowledge, did not write about his stay on the East Coast. However, his travels provided the perfect opportunity for him to learn more about these northern countries, including their specific cultures and the major changes they went through during the second half of the nineteenth century. While in London over the course of summer 1868, Husayn tried to learn English. He acknowledged wanting to master "the basics of their language for translating their books. As for speaking it, it is very difficult for one who is no longer young."[48] He knew Turkish, though not as well as Arabic; the inventory of his documents lists a limited number of letters in Ottoman Turkish.[49] He had a more active and creative knowledge of French and Italian. On occasion, he even dictated a few lines in French.[50] Between 1856 and 1872, he Arabicized several French and Italian terms, such as calling cards (*kārṭa al-fizīta*), pomade (*būmāda*), council of state (*kūnsīl dī Ātā*), protocol, director, tribunal, project, *aviso*, visit, and *camera* (chamber).[51] He also used, though less frequently, entire French phrases such as "il n'y a rien à perdre" ("there is nothing to lose").[52] When conversing during an encounter in September 1877 with Midhat Pacha, an Ottoman statesman then exiled in Europe, Husayn spoke to him in French.[53]

While in Europe, Husayn also played an active role in the book trade, patiently acquiring works in Arabic and expanding his library. And, like other Muslim travelers, he left an account of his European travels. The manuscript catalogued at the Tunisian National Library has been lost, but its title, *Travelogue of a Tunisian to Europe Written by General Husayn, Minister for Instruction in Tunis*,[54] indicates that it did not describe a voyage across the Muslim world, as had been the case for this genre in the past. Like other travelogues, or *riḥlāt*, composed by Husayn's North African and Ottoman contemporaries, the main purpose was no longer to describe the Muslim world and its wonders but to provide an account of Europe.

Most of the cultural production stemming from Husayn's travels was geared toward a Muslim audience—more precisely, his Tunisian friends and colleagues. Some of the books he collected were sent to Tunisian dignitaries.[55] As Husayn traveled across Europe and the Ottoman Empire searching for knowledge and awaiting a favorable change in the political winds, he constantly met with Muslim notables. In 1868, in Cairo, Husayn made the acquaintance of 'Abdallah Abu Sa'ud, the owner of the *Wādī al-Nīl* (*River Nile*), a weekly journal launched in 1867; and of Rifa'a Rafi' al-Tahtawi, an *imām* who, after traveling to France in 1826, had founded a language school and translation centers in Cairo.[56] In the same city, Husayn built up lasting business ties with the Moroccan Ibrahim al-Sanusi, whose brother 'Amr al-Sanusi was part of the inner circle of Shaykh Muhammad Zafir al-Madani, supreme head of the Madaniyya brotherhood.[57] Founded in the 1820s in the town of Misrata in Tripolitania, it had since spread across the surrounding territories towards Fezzan, Tunisia, Egypt, and Hejaz. In 1876, Zafir al-Madani's son and heir, Muhammad, settled in Istanbul, where he acquired considerable influence with Sultan Abdülhamid II, particularly in Egyptian and North African affairs.[58] Husayn consolidated these ties during his visits to Europe, taking his place in the world of dignitaries who moved between London, Paris, Rome, and more remote locales. For instance, after his stay in Cairo in 1868, Husayn, four of his compatriots, and two Algerians attended the funeral of an old Persian man who was buried in a small Muslim plot, close to that of the Jews, in London.[59]

Throughout these many meetings across Europe, and with the knowledge they provided, Husayn and his various acquaintances participated in a major intellectual dynamic in the second half of the nineteenth century. He and his counterparts in Europe debated the very meaning of the Muslim world. According to historian Nile Green, throughout their discussions, these Muslim travelers reconceptualized their understanding of the Islamic world and then exported their new interpretations back to Muslim lands.[60] However, unlike other Muslim travelers, Husayn did not discuss these topics with European orientalists. His own thinking on the boundaries between Europe and the Muslim world evolved as he extended his stay in Europe and slowly distanced himself from Tunis.

In 1873, Husayn was involved in what would be his last attempt to implement reform in Tunisia. His patron, Khayr al-Din, became the new prime minister, replacing Mustafa Khaznadar. Khayr al-Din formed a new government and tasked Husayn with supervising and reforming education in Tunisia.[61] But the government did not last. The Tunisian governor, Muhammad al-Sadiq, strongly disagreed with his prime minister's policy and ousted Khayr al-Din in 1877. In one of the many letters he sent to his patron, Husayn openly expressed his anxiety. In 1877, he felt he was witnessing the end of an era, as mamluks like himself, who had once held the reins of power and implemented reforms, were on the verge of disappearing. Husayn asserted that the "people's desire (*murād al-qawm*) is to be rid of mamluks" of his sort.[62] Free Tunisians were replacing the mamluk slaves. This was obvious to Husayn: a native Tunisian, Mustafa Ibn Isma'il, replaced the Circassian mamluk Khayr al-Din, who left Tunis for Istanbul in 1878 to become the sultan's main adviser, or vizier. Husayn, as we know, settled in Tuscany, and continued to express strong opposition to Muhammad al-Sadiq's government.

At this point in his life, less than a decade before he passed away, Husayn's devotion to reform had not disappeared. In fact, it had shaped him to his core: the way he understood himself, his world, and his ability to act were all informed by it. Husayn's exile in Tuscany was therefore not a distinct period of his life, a total separation from his Tunisian past. Indeed, his many voyages in Europe and within the Ottoman Empire since the 1850s had prepared him for such an exile. In Tuscany, as we shall see in the following section, Husayn adopted new narrative and writing forms and techniques. Above all, he reconceptualized the relationship between the Ottoman Muslims, European Christians, and North African Jews.

Exile as a Redefinition of the Self and the Muslim World

Even before the French conquest of Tunisia in 1881, Husayn considered himself an exile. Husayn labeled a trip in 1878 as a *hijra*. This term is crucial. It refers to an enforced departure and to a founding event in Islamic history: the migration of the prophet Muhammad from Mecca to Medina in 622, a date identified as the beginning of a new era and the start of the Islamic calendar. Husayn's *hijra* took him "to an infidel country (*bilād al-kufr*) until God . . . shall console the sorrow that besets Muslims in the East and in the West."[63] Five years earlier, Husayn had even described the mission of one of his employers in Italy as *ghurba*, meaning an emigration or an estrangement.[64]

The Tunisian state did not sentence Husayn and his exiled friends to a "temporary or permanent ban," nor did it confiscate their assets "in the wake of political or legal offences."[65] Nonetheless, Husayn felt that his adoptive country had rejected him.[66] The Tunisian state was no longer the one that he and his entourage had served for more than half a century. This state was increasingly under the influence of the French consul Théodore Roustan in Tunis and his ally, Prime Minister Mustafa Ibn Isma'il. Husayn was not the first exiled North African.[67] With the French colonization of the Ottoman provinces of Algiers and Tunis, an increasing number of Algerian and Tunisian families had left their homeland. But unlike Husayn, most of these refugees settled in other Ottoman regions—in Tripoli, in Egypt, in the *Bilād al-Shām* (Syria, Palestine), and in Istanbul.[68]

In Tuscany, Husayn attracted Tunisian and other Muslim visitors. Under his influence, the city of Florence became a political hub for North African and Ottoman reformists. Rustam—Husayn's close friend who was also a Circassian slave and Tunisian dignitary—spent five days in Tuscany in 1878 before settling in the Egyptian city of Alexandria.[69] Husayn employed Bu Hajib as his secretary in Florence for six years, from 1876 to 1882.[70] Another major Tunisian scholar and Bu Hajib's student, Bayram V, visited Husayn on three occasions, in 1875, 1881, and 1887.[71] The third important figure of Husayn's intellectual circle, Muhammad al-Sanusi, left Tunis shortly after Bayram V, in May–June 1882 (*rajab 1299*). Al-Sanusi joined Husayn before traveling to Istanbul, where, out of a total of two hundred to three hundred subjects of the Tunisian *beys*, ten to twenty were political exiles.[72] Their situation worsened with the French conquest of Tunis. In 1882, Husayn lost all his official positions and all mandate to represent the interests of the province of Tunis.[73] The French colonial officers saw Husayn as a major opponent whom they needed to silence. Three years after being removed from office, in 1885, Husayn pointed out all his sacrifices for the Tunisian government,[74] denouncing the new agents that the French hired as "shameless interceptors and pilferers who would replace the sovereign's will with their own."[75]

Husayn and his Tunisian entourage were forced to adapt to this new situation. Losing their state support (and funding), they had to rebuild their network in exile, well out of their comfort zone. They had to find other ways to communicate now that they were at a distance from one another. After 1881, Husayn, his Circassian friends Khayr al-Din and Rustam, his scribe Bu Hajib, and the scholars Bayram V and al-Sanusi could no longer count on Tunisian agents representing their country abroad. The treaties of 1881 and 1883 that established France's protectorate over Tunisia placed all Tunisian subjects under French authority, which meant that from then on, French embassies and consulates represented Tunisians abroad. On stepping down from their positions, Husayn and his circle of exiles avoided using the telegraph and official postal services, for fear of being spied on. They handed over their letters to trusted emissaries, or to members of their circle on their many peregrinations across the Mediterranean.[76] Husayn and his acquaintances' fear of surveillance and, more generally, the fact of exile reshaped Husayn's understanding of Europe and his sense of belonging to the Ottoman Empire and the Muslim world at large.[77]

In Tuscany, Husayn was still interested in discovering more about European culture. He acquired European books, dictionaries, novels such as Voltaire's *Candide*, and an essay praising Arab civilization that circulated widely amongst contemporaries, *La Civilisation des Arabes* by Gustave Le Bon, first published in Paris in 1884.[78] Yet, at the same time, Husayn distanced himself from Europe. In a memoir drawn up in 1880, he started to highlight the disdain Europeans felt for his people, their lack of interest in Muslim knowledge, and their eagerness to consider Arabs as barbarians. In the memoir, Husayn railed against the Italian courts, which did not study "our laws" or "our customs" because Muslims were deemed to be "small," "weak," "barbaric," and "ridiculous."[79] Husayn was no longer the young Ottoman officer who felt closer to the French emperor than to the colonized Algerians when visiting Algiers in 1861. More than two decades later, in 1882, he wrote a pamphlet with al-Sanusi, harshly criticizing French colonial authorities and their allies. The pamphlet—published anonymously—denounced the turpitudes of Husayn's enemy, Mustafa Ibn Isma'il, a favorite of the governor of Tunis, who had fallen under the influence of Roustan, the French consul there.[80]

Husayn was clearly advocating for Muslim unity under the tutelage of Sultan Abdülhamid II. He frequently displayed solidarity with his Muslim brethren, helping finance the legal defense of 'Urabi Pacha, an Egyptian army officer put on trial by British occupiers after his troops revolted and were defeated in 1882.[81] Two years later, Husayn funded the Arabic weekly *al-'Urwa al-Wuthqā* (*The Trustworthy Bond*), published from March to October 1884 by two major Islamic reformists and thinkers, Jamal al-Din al-Aghani and his disciple Muhammad 'Abduh.[82] Historian Arnold Green has shown that Tunisian leaders and scholars such as Muhammad al-Sanusi thought that *al-'Urwa al-Wuthqā* was an "anti-Western secret society allegedly founded" by Jamal al-Din al-Aghani, with "numerous branches throughout the Islamic world."[83] When Muhammad 'Abduh visited Tunis from December 6, 1884 to January 5, 1885, Tunisians—thinking they would become part of this secret society—contributed funds to the newspaper. But most of its articles had less to do with French colonialism in Tunisia than with British imperialism in the Muslim world. *Al-Urwa*'s authors advocated for Muslim anti-colonialism and pan-Islamic unity.[84]

A New Attitude Toward North African Jews

While Husayn criticized Europe and vowed to act in favor of all Muslims across the world, he distanced himself from North African Jews, who, with the advent of colonialism, were caught between Muslim societies and the Western world. While in exile, Husayn started to become fascinated with European anti-Semitic ideas. In 1886, when he still had enough strength to write or dictate his letters, he often railed against the Jews who "have eaten us" (*wa naḥnu akalat-nā-nā al-Yahūd*)—"us" meaning the Muslims.[85] A few months before he passed away, the disgraced Husayn had become absolutely convinced that the Jews were the main cause of his misfortune. He rained insults upon his former secretary, French-Algerian Jew Léon Elmilik/al-Liyahu al-Malih, that "dog," "that Jew" who "has stolen [my] assets."[86] In the same letter from 1887, Husayn informed his protector Khayr al-Din that he was about to send him a copy of the French anti-Semitic pamphlet *La France juive*, published the previous year by Edouard Drumont. Husayn wrote that "that there were more [copies of *La France juive*] in Istanbul than in Italy," and he urged his patron to read it.[87]

What could explain Husayn's conversion to anti-Semitic ideas during his exile? In Tunis and then in Tuscany, from the 1860s to the 1880s, Husayn frequently worked with North African and Italian Jews, hiring them as accountants or translators. But Husayn did not consider Tunisian Jews his equals. In 1860, in his letter about the political rights of Tunisian Jews, he argued that none of the Jewish notables he knew in Tunis had the skills to join the Supreme Court. Husayn stated that Tunisian Jews had to change their mind-set and become loyal to the Tunisian state.[88] The French colonization of Tunisia confirmed his perception of Tunisian and North African Jews. Most of his Jewish employees—the accountants, mediators, and translators hired in Tuscany and Tunis alike—turned their backs on him, siding with the French colonizers. His right-hand man, Léon Elmilik, purloined thirteen sets of documents from Husayn in 1885 and sold them to the Parisian Erlanger Bank, which had granted a loan to the Tunisian state in 1863.[89] A vigorous legal tussle ensued between Husayn and his own representatives in Italy and Tunis throughout 1886–87.[90]

Husayn felt that the world he knew, the world he aimed to reform, had been turned upside down. He used to give orders to "his" Jews; now Jews like Elmilik were raising their own claims and no longer accepting Muslim supremacy. In Tunis and elsewhere, other North Africans had the same impression: European imperialism, especially French colonialism, was upsetting the formal hierarchies between Muslims and Jews. In colonial Algeria, the Crémieux decree gave Algerian Jews such as Elmilik full

French citizenship in 1870, while most Algerian Muslims were deprived of basic political rights until the Second World War. In colonial Tunisia, the local Jewish minority was never collectively granted French citizenship. However, as in Algeria, an increasing number of Jews enjoyed better access to French schooling and therefore better social mobility during the colonial era.[91]

But Husayn's anger and his frayed relationship with his Jewish employees were not due only to changes in the balance of power between Jews and Muslims in Algeria and Tunisia. Husayn's exile in Tuscany also deeply influenced the ways in which he expressed his rancor toward Jews. Husayn echoed anti-Semitic opinions voiced by his Christian staff. One of his servants, Eva Keusch, the French-German mother of one of the two girls Husayn raised in Tuscany, openly expressed anti-Semitic sentiments. After Husayn's death, she spoke out angrily against the institution representing France in Tunis (the *résidence générale*), which only "serves to give jobs to the Jews."[92]

Drumont's anti-Semitic *France juive* seemed to mirror Husayn's frustrations. As shown by Dorian Bell in *Globalizing Race*, Drumont invented "a synthetic new anti-Semitism" by exploiting "France's discovery of North African Jewry, along with metropolitan financial scandals that negatively implicated Jews in the rapid colonial expansion of the 1880s."[93] Not only did Drumont denounce the Crémieux decree that granted full citizenship to Algerian Jews, he also argued that Jewish holders of Tunisian debt—such as the Erlanger Bank—had ordered France's 1881 invasion of Tunisia. This hugely distorted version of a complicated historical process aligned with Husayn's understanding of his own experience of losing political influence in 1878, as he could never admit that the reforms he and others had implemented over the course of the nineteenth century could not prevent France's takeover of Tunisia.[94]

A Hidden Lineage of Women

Husayn's exile in Italy had one last, even more unexpected outcome: by the mid-1870s, Husayn had established an atypical lineage. He was nominally married to Hanani, Khayr al-Din's daughter, who lived in Istanbul with her father and family from 1878 on. At the same time, Husayn was raising two young girls in Tuscany. Both children had two first names. The first girl's Christian name, Emma, sounded like her other name, Amina, meaning the "faithful" or "trustworthy" woman in Arabic. Emma/Amina was born in Livorno on March 1, 1876,[95] but it took more than four years for her parents, Eva Keusch and Sadiq al-Mu'addib, to recognize her as their daughter in front of a notary on November 13, 1880.[96] The father, Sadiq al-Mu'addib, was one of Husayn's adjutants in Tuscany.[97] Sadiq's father, Shaykh Ahmad al-Mu'addib, was a highly regarded Muslim scholar in Tunis, a jurisconsult (*faqīh*) and a teacher (hence the family name *mu'addib*). In Tunis, Sadiq already had three children (Hamida, Fatuma, and Duja) with a colonel's daughter. Though he did little to look after Emma/Amina, he did not forget her when he went back to Tunis.[98] On December 23, 1880, one month after the declaration before the notary, he sent the child a letter with an item of clothing.[99] Sadiq died before Husayn.[100]

Emma/Amina's mother, Eva Keusch, was born in the Grand Duchy of Baden, a state in what is now southwest Germany neighboring French Alsace.[101] In 1853, Eva's parents settled in colonized Algeria after having obtained an immigration permit and renounced their German nationality for French citizenship.[102] Eva later moved to Tuscany, where she was employed for a while as a housekeeper in Husayn's household. She became pregnant while working there. In 1877, according to Eva, Husayn adopted Emma/Amina "with the intention of being able to marry her off."[103] Two years later, in a deed preceding Eva's departure for a new life in North America, Eva agreed that her daughter would "remain with General Husseïn"; on the back of the deed, two words in Arabic stand out: "mother" and "daughter."[104] In 1880, Eva married Bertrand Meuville, a New York businessman living in Montreal,[105] but she did not forget Emma/Amina and continued sending her letters during the 1880s.[106]

The second child raised by Husayn in Tuscany had two first names as well: a Christian name, Maria or Marianina (meaning "little Maria" in Italian), and Myriam, the Muslim version of the Virgin's name. Like Eva Keusch, Maria/Myriam's mother, Angiolina Fortunata Bertucci, was an employee in Husayn's household—one of the Tuscan servants sought after "for their cleanliness" by nineteenth-century elites in the south of France and later in colonial North Africa.[107] Husayn did not officially recognize Maria/Myriam's paternity, and there is no record of the 1884 birth in the Livorno civil register under the name of Angiolina Bertucci or that of Husayn Ibn

'Abdallah. However, Angiolina revealed the actual nature of her relationship to Husayn in an undated letter written in broken French:

> Dear General, I am very glad to hear that you are feeling better and better. . . . However, I hope to see you again soon. The other day, Marianina saw the wet nurse ironing your flannel, she recognized it and hugged and kissed it, saying my dear daddy, come Daddy come. Receive one thousand kisses from Amina and Marianina, and one thousand affectionate greetings from your Most affectionate [Angiolina].[108]

Husayn acted as Angionlina's and the girls' protector. He sent biscuits and chocolate from fine Florentine confectioners.[109] His letters conveyed his deep affection for the two girls. In an 1884 diary entry, Husayn referred to Maria/Myriam as "my daughter."[110] He also kept notes from Emma/Amina, particularly a letter she sent him when she was about ten.[111]

Husayn's case was not unusual among the exiled elites. Former Tunisian statesmen had to adapt their family customs and their conceptions of kinship in exile. When leaving Tunis in the mid-1880s, Husayn's friend Bayram V brought his "harem"—to use the French colonial administration's expression for his wives and children—to Cairo.[112] Other Ottoman dignitaries had mistresses and indulged in new affective and sexual experiences. Such was the case with Turkhan Bey, the Ottoman ambassador to Rome from 1877 to 1880, who would boast to Husayn in some detail about his love affairs, including his liaison with a Russian countess: "Any love for the countess was long ago thrown into the *bâkâl torbassi* (grocer's purse). Do not go accusing me of inconstancy, as I am not to blame for our separation. For as long as a woman is unsure of her victory, she struggles. . . . But I pity the poor woman, not because we have fallen out, but because she is in a state of poverty to which she is unaccustomed."[113] Rustam, another of Husayn's closest friends, a former Circassian slave and Tunisian dignitary, was also married to one of Khayr al-Din's daughters. However, at the end of his life, in Alexandria, Rustam took an odalisque (*jāriya*) he had received as a gift for his concubine.[114] Even Husayn's enemy, former prime minister Mustafa Ibn Isma'il, fell in love with a certain Marie Abadie in Paris in 1887; Marie followed Mustafa to Tunis and eventually became his third wife.[115] But Husayn went beyond love stories and sexual companionship. In his will, Husayn clearly states that he raised the two girls in the Muslim religion.[116] More importantly, in the same will, Husayn expresses his intention to bequeath a third of his goods to Emma/Amina and Maria/Myriam, provided they marry Muslims on coming of age.[117] According to Islamic law, a Muslim could bequeath one-third of his property to people who were not his legal heirs without the consent of those heirs. But why did Husayn make these choices? Why did he create a special bond with these two girls over the last decade of his life?

Until he passed away, Husayn cared first and foremost about social reputation. When he decided to raise Emma/Amina in 1877, the same year as his departure from Tunisia and the Tunisian government, he wanted to protect the reputation of Emma's father, Sadiq al-Mu'addib, who already had a wife and children in Tunis and who belonged to a respected family. Husayn's keeping of Emma/Amina avoided a major scandal: Eva Keusch made no claim against Sadiq, and the Mu'addib were still seen as a virtuous family in Tunis. When Husayn subsequently took care of Maria/Myriam, he also took care to separate his public and private lives. In Tuscany, he divided his time between multiple residences, one to the north of Florence (at via Fra Bartolomeo 40), a second in the surrounding countryside (the villa Sicard), and a third in Livorno.

For Husayn, Florence was a place of social and professional interaction, where he received exiled Tunisian guests. Livorno was a family place, where he kept the two girls and one of their mothers. In his multiple letters to his father-in-law Khayr al-Din, Husayn kept nothing back about his emotions, his professional frustrations, and his physical pains. But in none of them did he ever mention Angiolina Bertucci, Emma/Amina, or Maria/Myriam. Only a few people in Husayn's Tunisian circle were aware of their existence. When writing his will in 1887, Husayn consulted Bayram V about what share of an inheritance could be left to the two daughters. Husayn's secretary, Shaykh Bu Hajib, who lived in Florence from 1876 to 1882, and 'Amr, the Shaykh's elder son who helped Husayn with daily life, clearly knew Angiolina and the daughters. In his will, Husayn granted 'Amr a role similar to that of a tutor under Muslim law, being in charge of protecting the two girls and their mothers. 'Amr even married Angiolina after moving from Livorno to Tunis in 1888.

Initially, Husayn could not formalize his relationship with Angiolina and the two girls for several reasons. First, he was already married to Hanani, Khayr al-Din's daughter. Second, Angiolina was not a Muslim. She was a Christian servant from the lower social classes. Third, the practice of formal adoption was not typically allowed: the "earliest Muslim jurists" classified it as "a prohibited act."[118] Above all, Husayn might have thought he was contradicting the political commitments that he had embraced since 1881: he had an Italian concubine while criticizing and fighting against European imperialism. He refused to live in Istanbul with his official wife and father-in-law even while constantly advocating for Muslim unity under the aegis of Ottoman Sultan Abdülhamid II.

Husayn only resolved these apparent contradictions during the two last years of life. While formally remaining loyal to Khayr al-Din until the end of his life, on September 6, 1886, Husayn repudiated Khayr al-Din's daughter Hanani, with whom he had never lived.[119] Such repudiation had a significant legal consequence: Hanani and Khayr al-Din could no longer inherit from Husayn. Through this repudiation, Husayn avoided the fate of his friend Rustam, the Tunisian dignitary and former Circassian mamluk, who was still married to Khayr al-Din's eldest daughter when he passed away in Alexandria and who therefore left all his wealth to Khayr al-Din's family.[120] Only a few months after repudiating Hanani, Husayn decided to pass a third of his estate to Emma/Amina and Maria/Myriam, as well as to bequeath some lands in Tunisia to Shaykh Bu Hajib and his sons. Husayn was not only acting out of affection. By the end of his life, Husayn was an emancipated slave who was no longer married and had no recognized descendants. In such a situation, according to Islamic law, Husayn could only bequeath his estate to his former master and his family—that is, to the Ottoman governor of Tunisia and his dynasty—even though Husayn had repeatedly criticized them and distanced himself from them since the late 1870s.[121]

In order to contest the Tunisian governor's authority, Husayn availed himself of other options that were legal in Islam: he named the Bu Hajibs administrators of his Tunisian lands and granted a third of his wealth to Emma/Amina and Maria/Myriam, who were not officially his legal heirs. To bequeath his estate, Husayn used methods of establishing a lineage that he was familiar with from his own childhood. Just as he had grown up in the Bardo Palace and been educated alongside the children of the governors of Tunis, he raised Emma/Amna alongside Maria/Myriam in a Tuscan house in Livorno. In the same way that Tunisian governors considered Husayn and other mamluk dignitaries to be their "true" sons without adopting them, Husayn referred to Myriam as his daughter in private letters, without acknowledging this status in official deeds. To take this comparison a bit further, Husayn might even have considered his German and Italian employees, Eva Keusch and Angiolina Bertucci, to be the concubines in charge of his household, able to procreate for him and other male members of his household.

Husayn's first experiences of kinship in the Tunisian governors' palace while he was still a child affected the last episode of his life, as he was writing his will. Even the ways in which Husayn shaped a lineage that would allow him to bequeath his inheritance as he pleased during the colonial period is reminiscent of Ottoman notables' household politics during the early modern period, when the notables acted and competed through their households across the empire, and when kinship and politics were closely intertwined.[122] Husayn's life, and surely the lives of many other North Africans, should therefore be viewed not solely through the lens of colonialism or even North African pre-nationalism but also through an Ottoman political culture that still permeated some sectors of North African societies. In this chapter, as I followed Husayn's trajectory, I have argued that "precolonial" experiences, and above all Husayn's sustained involvement in implementing Ottoman reforms, had long-lasting effects. These experiences, and the knowledge he gained from them, were still at play during the colonial era, or at least during the first decades of the French administration of Tunisia.

Clearly, French colonization affected Husayn deeply. His health deteriorated following the French invasion. His perception of Christians, Muslims, and Jews became far more antagonistic. In the end, Husayn no longer felt he belonged to the same political world that had brought together the Ottoman Empire and other imperial regimes in Europe. He no longer thought that Tunisian Jews had to be loyal to their state; he rejected the Jews, pure and simple. But most of these changes in Husayn's outlook, and perhaps even his health, were initiated before the French conquest in 1881. Most of Husayn's political attitudes in exile were related to the Ottoman reforms and their legacy. In the Ottoman province of Tunis, the governor stopped invoking the language and mind-set of reform in the mid-1860s, when he revoked the 1861 constitution, and then again in the mid-1870s, when the governor ousted Khayr al-Din from his post as prime minister. Each time, Husayn felt betrayed and frustrated. Each time, he distanced himself from those whom he considered to be the enemies of his country and of the Muslim world.

Husayn's life can therefore only make sense in a context that transcends the dichotomy of the colonial and "precolonial" eras. Husayn's understanding of slavery, for instance, shows that its abolition was not brought about only through the confrontation between Western abolitionists and Muslim societies. Elite slaves like Husayn themselves played a role in the abolition of slavery. By the same token, Husayn's involvement in implementing Ottoman reforms was not only a political struggle to strengthen and centralize the state administration to resist Western imperialism. Such reforms also went hand in hand with new understandings of writing techniques and publication technology, such as the newspaper and the legal pamphlet. Finally, Husayn's deliberate creation of a lineage to ensure transmission of his legacy outside the bounds of colonial occupation sheds light on the transformations of lineage and households in exile during the colonial period. Historians of Europe, along with those of the Ottoman Empire and the Muslim world, have shown that various types of families existed throughout the early modern period.[123] But there remains much to discover about how Muslim families (or those from the Muslim world) adapted to exile and how they passed on assets despite the constraints of exile and colonial domination.[124]

TWO

Husayn's Wealth

How to Build and Protect an Estate Between Empires

A FEW books in Arabic and in French, some suitcases, two horses purchased in Tuscany. . . . When Husayn passed away on June 27, 1887, he left very few belongings in his two Florentine apartments. The Italian authorities auctioned off these items nearly five and a half months after his death for the modest sum of 4,500 French francs, hardly anything compared to his total assets of 1,500,000 francs invested outside of Italy.[1] Most of Husayn's estate was located in Tunis and the surrounding area. Husayn began managing buildings and lands there in the 1860s when he rose to prominence in the Tunisian administration. In addition to two modest properties, he owned a large residence in the old city, or medina, of Tunis, called Dar Husayn (Husayn's house or palace).[2] He also held nearly twenty hectares of land in the expanding strategic port of La Goulette (thereafter referred to by its Arabic name Halq al-Wadi). Additionally, beginning in the late 1870s, he was involved in a Mediterranean network of creditors: some North African Jews provided loans to Husayn in the Tuscan city port of Livorno, and Husayn in turn lent sums to his exiled Tunisian friends and acquaintances on the southern shores of the Mediterranean.

These assets, loans, and various belongings will be at the center of this chapter. While the first chapter explored Husayn's life and his extensive involvement in Ottoman reforms, this second chapter will focus on yet another significant aspect of Husayn's life and afterlife: the *fin de siècle* material culture Husayn took as his own, his estate, and his moral understanding of economic transactions. I will show that the various dimensions of Husayn's estate can open up new understandings of material culture, property, and ownership in the period of transition between Ottoman rule and the French colonial era in Tunisia and surrounding North African countries. While the history of property and ownership during the colonial period was characterized by violent dispossession of the colonized subjects by European states and settlers, in this chapter I argue that—in the Tunisian case at least—we need to understand this history of dispossession within a broader context of land reforms preceding the 1881 colonial conquest, beginning in the mid-nineteenth century. At that time, Ottoman dignitaries such as Husayn pushed for a redefinition of public and private property. The fact that Ottoman dignitaries were deeply involved in implementing these land reforms is crucial for our understanding of the colonial period in Tunisia. Husayn built on his administrative knowledge and his personal experience of land and real estate management acquired during the reform era to protect his wealth until the first years of the colonial period and to pass on his real estate in 1887. Indeed, not all colonized North African subjects were as experienced as Husayn in protecting their assets. But I contend that to better understand the many effects of the land dispossession in colonized Tunisia, we need first to study the practices, types of knowledge, and Maghrebi values related to property and ownership before foreign occupation, since all of them were still used and adapted during the colonial period.

In the first section, I will explore Husayn's relationship with Ottoman and European material cultures from the mid-nineteenth century to 1887. I will argue that Husayn had no problem adapting to European bourgeois material culture and easily managed to amalgamate this culture with a North African–Ottoman way of life during the colonial period. Since the 1840s and his schooling at Bardo, Husayn often combined European outfits with Ottoman headgear (the *shāshiyya*, a red hat produced by Tunisian craftsmen).[3] He was accustomed to various European and North African domestic ways of life through experiences ranging from his Tunisian training to his Florentine exile. By the end of his life, he had transitioned from military uniforms to civilian clothing while still caring for and defending his Tunisian real estate. Even when threatened or dispossessed, Husayn (and probably other colonized subjects) were not always torn between European and Muslim material cultures. What clearly mattered for Husayn, from the mid-nineteenth century to the end of his life, was his estate in Tunis and, more broadly, the issue of land ownership.

In the second section, I will explain the significance of land ownership at that time by placing the development of Husayn's estate within a broader transimperial context. Throughout the nineteenth century, in rural North African societies as well as in Europe, various understandings and practices of land ownership coexisted and overlapped to a certain extent. Private property was not the exclusive, or even the main, form of ownership. State ownership and collective holdings were still significant on both sides of the Mediterranean.[4] However, beginning in the 1830s in French Algeria, the colonial state started to privatize property as the French army and European settlers expelled colonized Muslim subjects from their lands. By the end of the 1850s, across the Ottoman Empire, from the Tunisian province to the Middle East, new land regulations promoted private ownership. Through this new legal framework, Ottoman dignitaries were pursuing two main goals: on the one hand, the state elite was reinforcing its domination over land and agriculture, which was the main source of economic wealth across the empire; on the other, that same elite aimed to control European traders' and companies' legal access to strategic land and buildings. By closely studying the composition and management of Husayn's real estate in Halq al-Wadi, I will show that Husayn and the Tunisian state elite participated in their own way in crucial transimperial debates that had significant consequences beyond the issue of property.[5] In Tunisia, men like Husayn combined public and private conceptions and practices of ownership from the 1860s onward to fund the urban development of Halq al-Wadi and to monitor land ownership despite increasing European interest and financial investment in Tunis and its surroundings.

The competition between local Muslims, Jews, and Europeans in Tunis and other parts of the Ottoman Empire became fiercer in the 1870s, as we shall see in the third and last section of this chapter. The Ottoman administration faced serious financial issues: the Tunisian state went bankrupt in 1867, and Cairo and Istanbul followed by the late 1870s. In this new economic context, Husayn and other landlords in Ottoman Tunis had to cope with the International Finance Commission, a new institution implemented in 1869 to manage Tunisian public debt. After that period, Husayn and his Tunisian entourage were no longer part of the same conversation with Europeans about private and public property. Husayn developed his own moral discourse about ownership, claiming that Tunisians could rent but should never sell land and buildings to Europeans. In the 1870s and 1880s, as he became involved in his network of creditors, Husayn became acquainted with the world of finance and gained a certain level of expertise that allowed him to escape French colonial power, to a degree, and to defend his assets and what he defined as Tunisian state interests.

A Shared Bourgeois Culture

Like other Ottoman dignitaries of his generation, Husayn easily adapted to an evolving material world throughout the second half of the nineteenth century. Until the 1850s, Husayn attended Tunisian government councils held on divans, with low tables and rugs. In the mosques, he was also part of teaching circles (*ḥalqa*), which would gather on plaited mats next to columns. But in the 1860s, Husayn would move to different settings, sitting on stiff chairs around high tables in new municipal, judicial, and legislative institutions.[6] During the same period, Husayn went from a world of folds and drapery to one of uniforms with sharp outlines. From the 1830s onward, Husayn had modeled his appearance on European and Ottoman military styles, with a red Tunisian hat (*shāshiyya*) on his head. In his dress, Husayn resembled other civilian and military administrators in Istanbul and its provinces, whose new state and military outfits were part and parcel of the Ottoman reforms that were being implemented. As argued by historian Charlotte Jirousek, "clothing was perceived to be a major implement for building the changes in values that were devised."[7] Military and administrative officers went therefore "from gender-neutral to gender specific norms,"[8] visually differentiated from Muslim scholars (*'ulamā'*) sporting turbans and from women, who were still wearing loose clothing.[9] In specific contexts, however, Husayn played with his various apparel options. For instance, in the picture by Félix Nadar on the cover of this book, we see that Husayn mixed items from multiple traditions for the specific needs of this staging: European shoes, shirt, and bow tie are paired with North African articles such as a turban, embroidered vest and jacket, and the famous baggy trousers called *sirwāl* in Arabic. The *sirwāl* was already an integral part of the military uniform of some North African regiments, including the *zouaves* during the colonial period and the mamluks escorting Napoleon at the beginning of the nineteenth century.

As of the 1870s, in exile in Florence, Husayn adapted to yet another material world. He no longer wore his Ottoman-Tunisian military uniform. Instead, he sported civilian garb, hats, and fur-collared jackets.[10] He decorated his house in the Italian style of late-nineteenth-century Tuscan residences. As in these other households, Husayn was surrounded with everyday objects such as cooking utensils, lounge and bedroom furniture, and a pianoforte;[11] and, as in other households, Husayn transformed certain daily objects into collections of items to be admired, including certain books and Etruscan-shaped amphora.[12] Other objects held sentimental value for him, such as Emma/Amina's clothing from when she was very young or children's reading manuals that he kept with him in Florence.[13] Husayn ran his home and business accounts in the European bourgeois fashion, except that he kept his records in Arabic, not Italian.[14] He was not trying to adopt European customs; he was merely the product of a blended society of Ottoman, Maghrebi, and European influences.

Husayn even played a key role in the development of an imperial bourgeois culture in the Mediterranean by responding to the desires of Tunisian courtly society for European goods.[15] He served as an intermediary, bringing items from Europe to Tunisian government circles. Husayn received orders for books, fabrics, and weapons from his peers. He sent two dogs by ship from London to his protector, Khayr al-Din.[16] In 1854, he purchased jewelry worth Fr.600,000 for Tunisian prime minister Mustafa Khaznadar.[17] Husayn provided similar assistance to Europeans, who brought back objects from their voyages and transformed them into what European cultural historian Krzysztof Pomian has called "semiophores"—objects collected "not for their usefulness but because of their meaning" and because they carried a certain social cachet.[18]

Husayn lived in Tuscany for more than fifteen years but owned very little there. Like other exiles and migrants who thought they were passing through, he rented furnished accommodations.[19] His apartments were full of sometimes worn pieces of furniture (including an old red armchair), bags, carryalls, and cases.[20] The only things he kept from Tunis were precious stones set in military decorations, which disappeared just before Husayn passed away.[21] Waiting outside the door were the horses he owned until he died, ready to take him on one of his many journeys.[22] His horses and personal effects were sold for Fr.4,500.[23] Husayn did not even seek to hand down his book collection. He was a keen bibliophile, but he did not set up a foundation (*ḥubūs)* for the books he had collected, as other Muslim scholars and dignitaries had done before him,[24] nor was he consumed by any desire to itemize the objects he acquired in Tuscany.[25]

Husayn certainly knew that his goods would be auctioned off to reimburse his creditors. Since this was the case, was it in his best interest for his possessions to be properly registered in Florence? More fundamentally, why should he own objects in exile? Owning, acquiring, and passing on property may well have granted or, conversely, limited his rights in Tuscany. But did men such as Husayn even seek recognition of these rights in their places of exile? In Florence, Husayn continued to assert his social importance. For instance, he would insist on his title of *pasha* (an honorary title granted to Ottoman grandees) in Italian newspapers and in his correspondence. In fact, Husayn drew clear distinctions between his limited material means in Florence, his social rank, and his wealth concentrated in Tunis.[26] Other Ottomans and North African colonized subjects from the middle class strove to distinguish themselves from the West, refusing to imitate European habits and fashion.[27] However, what mattered to Husayn was less this material culture than his palace in Tunis and his lands in Halq al-Wadi. More broadly, throughout the nineteenth century, what appeared to be instrumental for Ottoman–North African subjects like Husayn was land ownership in the Maghreb.

Private Property and Public Domains: A Transimperial Issue

In Ottoman North Africa, as in other parts of the Islamic empire, until the mid-nineteenth century, ownership did not stem from any clear-cut distinction between public domain and private property. As in Europe, private property was not the most widespread form of ownership.[28] Multiple legal claims on lands were "often attributed to multiple social actors."[29] From the sixteenth century through to the second half of the nineteenth century, the state, acting in the name of the Muslim community under the Ottoman sultans, was in charge of all lands.[30] Under this principle, the sovereign retained eminent dominion (*raqāba*) over what was known as *miri* land,[31] although the sultans conceded individuals rights to use lands (*taṣarruf*), often in perpetuity.[32]

In Ottoman Algiers and Tunis, state lands were not labelled *miri*; rather, they were known locally as *beylik*, meaning that they were under the control of local *beys* and *deys*, the local governors. Sharecroppers (*khammās*) worked these lands. Tribes could also communally work land called *'arsh*, which was "held individually, to the benefit of each individual family group."[33] A third type of land use existed in which some families—often from the elite—managed to assert individual rights to lands that were called *milk* ("possessed" property). These families either held notarial titles or benefited from a consensus within their own community that acknowledged their long-standing reputation as workers and holders of these specific lands.[34]

French colonization of Algeria was the first major breach in this system. Beginning with the occupation of Algiers in the 1830s, confiscation of lands and "the constitution of private property" became "the essential foundation on which colonial society" was built, as summarized by historian James McDougall.[35] Initially, in 1830, all the *miri* or *beylik* lands became part of the French state's domain. But in 1851, a land law "established the state's 'private' rights of property." Soon after, an 1863 law (the Second French Empire *sénatus-consulte*) extended from France to Algeria the parceling of common property into individual hands.[36] This policy not only aimed at commoditizing lands for the interests of European settlers; it also legalized the confiscation of Muslim lands.

Outside French Algeria, across the Ottoman Empire, the land regime started to change in 1858, with the concomitant promulgation of land regulations in Istanbul (*Arazı Kanunnamesi*) and Cairo.[37] The new laws referred to individual property and indicated procedures for recording land and other possessions.[38] They established state institutions as arbiters in charge of determining private ownership rights between competing interests.[39] In Egypt, alongside a "vast mapmaking and land registry enterprise" from 1892 to 1907, books continued to promote private ownership.[40] Historians do not agree on the rationale of these state regulations. Did Ottoman lawmakers believe that private property would endow peasants with rights over the land they held so that the peasants would be more productive and enrich state coffers?[41] Or was the promotion of private property yet another way to consolidate the predominance of the state (and hence of the governing elites) in land ownership?[42]

In fact, the enforcement of the 1858 land laws depended on local context and therefore had varying effects. In Ottoman Tripoli, "private ownership and registration were introduced in the settled areas of Tripolitania [in the western part of the province] and urban areas of Cyrenaica [eastern part of the province]"; there, according to Lisa Anderson, "tribal lands were divided."[43] But because agriculture was a "risky business" in this region, land reform did not result in the concentration of property.[44] In the neighboring province of Tunis, unlike in Cairo, Istanbul, or French Algeria, the local governors first implemented land-related regulations in the 1857 Security Covenant (*'Ahd al-Aman*). This fundamental law proclaimed the "civil and religious equality of all subjects." It terminated the "monopoly" of Muslim landowners and allowed Jews and Christians to become property owners.[45] European speculators and businessman had been flooding into the country along with poor Maltese and Italian workers since the mid-nineteenth century. For the first time, members of these religious minorities were legally entitled to own property in Tunis.

Additionally, the *beys* signed agreements with British and Italian powers redefining land access for Tunisians and Europeans. The Anglo-Tunisian agreement of 1863, like the Italo-Tunisian La Goulette treaty of 1868, guaranteed ownership rights to people from these two countries.[46] At the same time, urban expansion, increasing trade, and labor migration to Tunis resulted in the emergence of new classes of property owners.[47] Tunisian historian Abdelhamid Hénia argues that it was during this time that Muslim subjects started abandoning their urban properties in favor of acquiring land in the area around Tunis. The land they sought to wrest from rural dwellers became "the main source of revenue for Tunisian elites."[48]

Multiple Legal Claims to Husayn's Land

Looking closely at Husayn's main estates gives us a more concrete idea of how Tunisian governors operated on the ground when new regulations opened the land market to Jews and Christians. Husayn's specific case clearly demonstrates how Tunisian Ottoman state elites combined public and private rights over land to develop and protect their interests beginning in the 1860s. For instance, the two largest properties in Husayn's estate—the lands at Halq al-Wadi and his palace in the medina of Tunis (figure 2.1)—actually belonged to the Tunisian *bey* dynasty and to the Tunisian state more broadly. From time to time, these governors would pass on

FIGURE 2.1 Dar Husayn in Tunis (Nationaal Museum van Wereldculturen, TM-60028247)

some of their estates to their subjects, including their most trusted mamluks such as Husayn.

Husayn's palace in the medina was originally built on the site of a medieval castle. By the beginning of the seventeenth century, the palace hosted the official residence of a Turkish army leader, Dey 'Uthman, as well as the Turkish soldiers' or janissaries' council (*dīwān al-jund*), which governed Ottoman Tunis at that time. In the second half of the eighteenth century, the same palace was passed on to a mamluk of Georgian origin, Isma'il Kahiya, and then a mamluk and vizier of Moldavian origin, Yusuf Sahib al-Tabi. The palace would remain in the hands of Tunisian governors until the early 1860s, when it housed the administration of the municipality of Tunis under Husayn's supervision.[49] Six years later, Husayn claimed that the palace was under his care after Khayr al-Din accorded him the building.[50] In this case, Husayn privatized a state building that he then rented, but refused to sell, to the French army after the 1881 conquest.

Another part of Husayn's estate, the lands located in the port city of Halq al-Wadi/La Goulette, also initially belonged to the governors of Tunis. Governor Ahmad Bey bequeathed the lands to Greek mamluk Mustafa Khaznadar, his prime minister from the late 1830s to 1873. In 1861, Khaznadar gave some of these lands and their buildings to Husayn, who was then head of the municipality of Tunis.[51] At that time, the port of Halq al-Wadi was undergoing massive changes. Most of the 4,600 Italians originating from the neighboring islands of Sicilia and Sardinia flocked to this port city. British investors were planning to build a railroad linking the port to the main city of Tunis.[52] In this specific context of economic and demographic growth, Husayn established the lands he received from Khaznadar as a *ḥubūs* or *waqf*. According to Islamic law, this type of property was neither fully private nor public. It was set aside from the land market. Land committed to a *ḥubūs* or *waqf* could support the heirs of the *ḥubūs* founder. If the line of the descent ceased, the income from the *ḥubūs* would be used to support public and pious causes such as schools, mosques, or hospitals. For Husayn and other owners, committing land to a *ḥubūs* offered two main advantages. First, *ḥubūs* were not taxed. More importantly, no one—including the government—could confiscate the *ḥubūs*.[53] The *ḥubūs* was so advantageous that in 1881, according to historian Charles Harber, "about 25 percent of Tunisian land reposed" in *ḥubūs*, "including a large portion of the richest agricultural lands."[54] In the case of the Halq al-Wadi lands, Husayn then rented them, and the buildings on them, to Tunisians and Europeans alike.[55] Husayn, like other Tunis dignitaries, used a form of rental called *inzāl* (or *kirāʾ muʾabbad*), which was "a lease in perpetuity" on a *ḥubūs*.[56]

Husayn's lands were included in more comprehensive urban planning as well. On March 15, 1864, while the country was reeling from a tax revolt, the governor of Tunis, Muhammad al-Sadiq Basha Bey, issued a decree approving "the plan for a new town to rise up outside the fortified enclosure" at Halq al-Wadi.[57] As in Europe, the state undertook responsibility for development of the town. The new town was a partial replica of the new modern seafront at Livorno in Tuscany.[58] A certain Gauvat, a French engineer to the municipality of Tunis, devised the modernization plans. The squares bore the names of governors who had reigned over Tunisia during the nineteenth century, and the roads bore those of their chief servants (notably Mustafa

Khaznadar and Khayr al-Din).[59] The plan thus projected the governmental hierarchy onto a public space. As part of the state's urban planning, Halq al-Wadi lands clearly blurred the line between public and private. Former state lands became part of the Tunisian gubernatorial dynasty's estates before being shared with state servants, who then established them as foundations that could be neither bought nor sold. Tunisian and European tenants could, however, rent the buildings on these lands. The many-layered status of these lands thus enabled owners such as Husayn to develop these assets for their own interests and to mitigate the desires of Europeans and local people who were eager to appropriate these lands throughout the second half of the nineteenth century.

Acting as head of Tunis's municipal council, Husayn was in charge of carrying out the urban renovations as planned. He kept two original copies of this plan among his papers: Governor al-Sadiq Basha Bey signed one copy, and French engineer Gauvat the other. Husayn did not hesitate to defend his owner's rights. He sought rent as compensation from his master, the governor of Tunis, who erected barracks on Husayn's lands.[60] Similarly, as of the mid-1880s, Husayn demanded that the Tunis municipality pay rent (as an *inzāl*) for about a thousand square meters of land on which the municipality had built a market (figure 2.2).[61]

From the 1860s until his death, Husayn also personally managed his *inzālists*, the people renting his land in perpetuity. These tenants of lodgings, stores, and bakeries on Husayn's land at Halq al-Wadi all had to operate within a local Muslim and governmental legal framework. They were bound by the terms of the *inzāl* tenancies, the provisions set out in deeds establishing *ḥubūs* and the Tunisian governor's dominion over these lands. The *inzālists* were too diverse to unite against Husayn. As indicated in a table detailing the sums these tenants owed to Husayn between 1883 and 1889, fifty-five individuals and public entities (including a philharmonic society and the "Italian railways") held sixty *inzāl* tenancies on Husayn's properties, of which twelve were in Tunis and forty-eight at La Goulette.[62] Seven *inzālists* bore Muslim names, eighteen were Jews, thirteen were Italian, and nine were French. We see, then, that the *inzāl* tenancies on the Halq al-Wadi lands provided perpetual leasehold to Christians and Jews, two categories of the Tunisian population who, up until the early 1860s, were theoretically excluded from ownership. These *inzālists* were not standard tenants. They could not renegotiate their tenancies in the short term,[63] and problems

FIGURE 2.2 Main gateway to the market at Halq al-Wadi, located between what is now Avenue Franklin Roosevelt and Avenue Habib Bourguiba. Photograph by author.

with them were only rarely related to unpaid rent. The limited disputes arose primarily from subletting or sub-*inzāl*, which was not recognized by the Tunisian authorities.[64]

Except for this last breach in Husayn's control over his land in Halq al-Wadi, it seems clear that the management of this real estate and of the Tunisian palace accustomed Husayn to competing for property and monitoring his holdings. From the 1860s onward, for more than two decades, dignitaries like Husayn experimented with various legal tools, such as the *ḥubūs* and *inzāl*, that they were still using to protect their assets during the colonial period. However, by the end of the 1860s, and again at the end of the 1870s, landlords and notables such as Husayn had to compete with more powerful interests than just the *inzālists*. After a dramatic financial crisis in Tunis, they had to compete with European bankers and state representatives. Facing French, Italian, and British banks and governments, Husayn developed a defensive moral discourse and started to find ways to hide his assets from European control.

The Consequences of European Financial Encroachment in the 1870s

The Ottoman reforms and their enormous cost resulted in a financial crisis across the Ottoman Empire. With the implementation of legal and administrative reforms, the growing state apparatus rendered tax revenue insufficient. The Tunisian treasury could no longer count on the income from state monopolies that were abolished under British pressures in 1857. In 1863, the Tunisian state borrowed thirty-five million French francs from the Parisian Erlanger Bank "to meet an international debt of some 30 million francs."[65] Four years later, it became obvious that the state was unable to repay its 12-percent-interest loan—half of the state's normal annual income was already devoted to yearly payments on the loan.[66] Two years later, France, Britain, and Italy (the major European creditors of the Tunisian state's debt) pressured the Tunisian government into creating the International Finance Commission, which would manage Tunisian public finances.[67] The commission consisted of two highly ranked Tunisian civil servants, a French tax inspector (*inspecteur des finances*), and the European creditors' representatives.

Tunisia was not an isolated case, although it was the first Ottoman region to experience this financial transformation. By the mid-1870s, the Egyptian state endured a similar financial crisis, which resulted in the creation of a Franco-British consortium monitoring the national budget.[68] In Istanbul, Ottoman debt worsened after the Russo-Turkish War of 1877–78; as of 1881, Ottoman debt was under the scrutiny of a new Public Debt Administration consisting of bank representatives and European delegates.[69] Each of the institutions created in this financial turmoil—the Public Debt Administration in Istanbul, the Franco-British consortium in Cairo, and the International

Finance Commission in Tunis—allowed Europeans to control Ottoman revenue and state budgets. In Tunis, the International Finance Commission monitored tax reforms and the state budget for payment of the state's debt to such an extent that the commissioners would even intervene in Tunisian land management.

Husayn had to defend his rights to the Halq al-Wadi lands against the International Finance Commission. In 1875, he sought damages from the commission when a railway station was built, encroaching on his land.[70] Like the *inzālists* at Halq al-Wadi, the French deputy chairman of the commission, Edouard Le Blant, was forced to operate within the local Islamic legal framework. Le Blant requested a legal consultation (*fatwa*) from a Muslim adviser "to ascertain whether the places indicated on the plan" infringed on Husayn's property.[71] For this purpose, Le Blant approached a Muslim jurisconsult, the Maliki *bāsh muftī* al-Shadili Ibn Salih, who found in favor of the French administrator.[72] Le Blant argued that Husayn had "taken possession of a few meters of land in places which" had been assigned to public spaces in the initial development plan. Ironically, it was thus Husayn who had to reimburse the railway company.[73]

Husayn sought a second opinion from another jurisconsult, Shaykh Ahmad al-Wartatani,[74] who argued that "the persons who received" the property "may administer these places as they see fit."[75] Khayr al-Din, prime minister at that time, prompted Husayn to produce the original urban development plans for Halq al-Wadi. Despite swearing he would do so, Husayn kept the plans in his safe.[76] Notably, the Tunisian government did not issue a ruling in favor of Husayn. The International Finance Commission clearly changed the local conversation about land ownership. Tunisian statesmen such as Husayn and Khayr al-Din could no longer act among themselves to promote their own interests or what they perceived to be the Tunisian state's interests. They had to take into account, or even argue against, the commissioners' views.

It was at this moment of increased competition over lands and state revenue that Husayn began to adopt a moral view of Tunisian landholding. In the late 1860s, Husayn stated that he did not want to become wealthy by owning land in Tunis. In his letters to Khayr al-Din, Husayn claimed that he "came naked into the world" and "he would leave it as he had appeared."[77] He had "only lived in one of the rooms" in the palace in the medina out of "modesty," sleeping "on a camp bed, cutting meals short, and leaving the table before service ended, being unable to accept the series of dishes, which were too many for his taste."[78] Above all, by the end of the 1870s, Husayn argued that Tunisian people and their elites should not sell their properties to Europeans even if it was legal. Although he rented to Europeans, Husayn was hostile to the idea of Europeans, and especially the French, acquiring outright ownership in Tunisia.

Around the early 1880s, Husayn accused Prime Minister Mustafa b. Isma'il of having "sold the country to the French."[79] Along the same lines, Husayn did not understand why Khayr al-Din decided to sell a 100,000-hectare parcel located in Enfidha, between Tunis and the city of Sousse, to the French Société Marseillaise de Crédit (Marseillaise Credit Company). Khayr al-Din had received this estate from the governor of Tunis as a reward after a successful political mission in Istanbul. Husayn believed that selling the Enfidha land to foreigners, and especially the French, "will reinforce their power over us"; it would be better to "sell to people from the country (*al-ahālī*) or else to Italians," who had less power over Tunis than the French.[80] Dissatisfied with the sale, Husayn did not shrink from telling Khayr al-Din that he had lost "the popularity and affection" he had acquired in Tunis "over more than forty years . . . at a single stroke."[81]

During his exile in Tuscany in the late 1870s, Husayn went further than strongly worded letters. He started to shelter parts of his estate from French and Tunisian interference, including funding that he used to help exiled Tunisians across the Mediterranean or for underground action against French colonial domination.

Husayn's Deliberate Efforts to Pass on and Protect His Assets

Husayn's strategy becomes clearer when we examine his lines of credit. If we rely only on the first official French assessment of Husayn's estate after he passed away in 1887, it appears that Husayn's liabilities were very limited in comparison to his assets. The French administration in Tunis estimated that his debts stood at Fr.175,605, representing between a sixth and a fifth of his estimated assets of Fr.1,195,000 (consisting mainly of the palace in Tunis and the lands in Halq al-Wadi). At first glance, General Husayn's exile could explain his financial situation, as Husayn simply did not spend much while in Tuscany. However, this financial configuration—and, more specifically,

Table 2.1 Husayn's creditors in Tuscany

Creditors	Asserted credit	Altered credit
Spezzafumo	50,000	40,000
Eugène Rosa	30,000	22,500
James Rosa	18,105	18,105
Cardoso (bills signed for Husayn)	50,000	50,000
Moles (Molco)	8,500	8,600
Bertucci	50,000 (liability "deemed unfounded")	45,000 (Moreno)

the very limited liabilities in comparison to Husayn's assets as a whole—is still intriguing. Husayn's limited debts and credits greatly contrast with the insolvent estates often found in major Ottoman cities, and with the North African economic situation (particularly colonial Algerian and Tunisian debt) at the time.[82]

In fact, the balance sheet of Husayn's estate drawn up by the French administration is incomplete (table 2.1). It mentions only six creditors (Spezzafumo, Rosa, Cardoso, Moles/Molco, and Bertucci). Only one of these was on intimate terms with Husayn—namely, his former domestic employee Angiolina Bertucci, the mother of Maria/Myriam. In this specific case, the French administration judged the alleged debt to be of dubious legitimacy. The five other creditors were business relations who acted as intermediaries for Husayn and the Tunisian state during his missions in Tuscany and who did not hesitate to file their claims.

French authorities mainly paid attention to Husayn's debts when creditors raised their claims to the French consul in Florence. French colonial power was less informed on Husayn's financial engagements across the Mediterranean. But if we look closely at Husayn's letters to Khayr al-Din, it appears that Husayn lent money to his companions in exile, particularly to Rustam, in Alexandria, and to the Sanusi family, in Cairo. This Moroccan family received Fr.80,000 from Husayn, who had a business relationship with one of its members, Ibrahim al-Sanusi.[83] It is quite possible, then, that Husayn's assets and liabilities were even more significant than our current estimates, and that he probably chose to lend and give funds to friends in order to hide the money from the Tunisian and French states.

More evidence of Husayn's deliberate efforts to protect his assets can be found in yet another aspect of his estate. The sum total of Husayn's debts, Fr.175,605 (in fact Fr.217,605, including sums paid just after his death), corresponded almost exactly to what the Tunisian state owed Husayn for his services, Fr.200,000.[84] This was not at all coincidental. Husayn had a very clear idea of his accounts and how he should present them to French and Tunisian representatives. Husayn's parsimonious way of life in Tuscany and his vigorous defense of his Tunisian property rights prove that Husayn and his acquaintances had a very clear understanding of what he owned and could pass on, given the context of the colonial occupation of Tunisia and his situation in exile. As we saw in the first chapter, as an emancipated slave who had no recognized descendants, Husayn cautiously protected his assets from being passed to his official heir, the governor of Tunis, at that time 'Ali Bey III. Instead, the governor mainly inherited Husayn's debts and would have to repay Husayn's creditors. 'Ali Bey III and his family only managed to sell Husayn's palace to the French army in 1890, eight years after Husayn signed a lease with the occupying forces.[85] Not including the palace, Husayn decided to pass on a third of his estate to Emma/Amina and Maria/Myriam, the two girls he raised in Tuscany.

More importantly, well before registering a will, Husayn laid the grounds for handing down his properties in the port city of Halq al-Wadi. As explained earlier, he transformed his land there into *ḥubūs*, meaning that these properties' incomes could revert to private individuals (*ahlī*) and, later, to pious works of public utility (*khayrī*). In 1886, one year before passing away, he decided to appoint the sons of his secretary Salim Bu Hajib as administrators and medium-term beneficiaries of this *ḥubūs*. Husayn sought to recompense Bu Hajib's sons for their help during his exile in Italy.[86] He also designated invalid soldiers from the Ottoman army as the long-term beneficiaries of his *ḥubūs*.[87] Husayn made the same choice as the majority of *ḥubūs* founders in Tunisia: he decided that his properties' income would benefit individuals first.[88]

As long as Bu Hajib's sons were alive, the income from Halq al-Wadi would not fully fund public institutions. As a result, the Tunisian government could not consider Husayn's *ḥubūs* "public," and a specific administration, the *jam'iyya al-ḥubūs* in charge of public *ḥubūs*, could not take over management of Husayn's lands.[89] At the same time that Husayn was bequeathing his lands, the French colonial administration was very carefully and slowly reforming Tunisian *ḥubūs* and the *inzāl* lease in perpetuity on *ḥubūs* properties. The act of 1888 put this type of lease "up for public auction." In 1898, another decree "recognized the exchange of land placed in *ḥubūs* for money or other land."[90] Subsequent colonial laws dealt with the so-called public *ḥubūs* and not the ones that individuals such as the Bu Hajib sons would administer.[91] Husayn and the Bu Hajib sons therefore succeeded in protecting a core component of Husayn's estate.

Was Husayn successful in protecting his estate? Only partially—he mainly managed to pass on the revenue from Halq al-Wadi to the Bu Hajib family for a single generation. However, more than simply assessing Husayn's success or failure, in this second chapter I have argued that if we pay attention to land reforms in the second half of the nineteenth century, we can understand how individuals like Husayn accumulated knowledge and experience and developed moral ideals about land ownership, and how they still employed such expertise after the French occupation of Tunisia. Husayn's case concretely demonstrates how Tunisian dignitaries manipulated Islamic legal tools such as the *ḥubūs* and *inzāl* to control the port city of Halq al-Wadi, including during the colonial period, and how, after the 1867 Tunisian bankruptcy, financial knowledge became even more crucial to this state elite.

But is this story specific to Husayn and other Tunisian elites? The land laws implemented in Tunis after 1857 actually had significant consequences for the whole of Tunisian society: Jews and Christians were allowed to purchase land for the first time, which might have shocked some part of the Muslim populations. As we will see in the following chapter, these land ownership and financial issues triggered an increasing number of legal conflicts in precolonial Tunisia. Moreover, Tunisians were not the only Ottoman subjects to experience this massive change or to build on their experience during the colonial period. In the neighboring province of Tripoli (*Tarāblus al-Gharb*) and across the empire, a new legal framework redefined public domain and promoted private property as European migrants and investors were settling on the southern shores of the Mediterranean.

In this context, Algerian Muslims could be seen as the only exception to this major transformation. After the French invasion of 1830, they barely had room to maneuver, let alone any legal ability to hold onto their lands. Historian James McDougall reminds us that "by the First World War, Algerian peasants had lost the ownership or use of almost 11.6 million hectares of farm, pasture, and other land, including forests."[92] A middle stratum of Algerian peasant landowners participated in a slow reconsolidation of Muslim rural property, but from 1898 to 1920, "Algerian peasants still sold more land to Europeans than they bought back from them."[93]

Learning from one's competitors or keeping up with one's enemies did not mean besting them at their own game. Tunisian state elites were held back by their own interests from going too far in redefining public domain and the importance of private property. Dignitaries from Tunis, such as Husayn, Khayr al-Din, and Khaznadar, had to defend the estates they received from their masters, the Tunisian governors. It would not be in their interests to draw too sharp a distinction between the figure of the sovereign and the constitution of public domain, given that their social promotion was rooted in their personal, dependent relationships with the governor of Tunisia and in their legal status as slaves and then as emancipated slaves. Public domain was therefore not defined in Tunisian law until the French authorities did so in 1885.[94]

By the late 1850s, Husayn and his entourage were involved in a transimperial debate about public domain and private property. But a decade later, when a financial crisis led to the inability of the Tunisian government to control its own budget, Tunisian statesmen no longer had a voice nor did they hold a financial stake in the new capitalistic trend that was clearly threatening their own estates and power. Husayn may have shared a material bourgeois culture with Europeans, but he also diverged from them in important respects. He started to develop a moralistic discourse of land ownership based on national origin, and he began to hide his assets from the Tunisian state and the French colonial administration.

This "great divergence" between local Ottoman dignitaries and Western capitalists in the late 1860s had another major consequence:[95] Husayn was not interested in investing in Europe. His major concern was to protect his assets in Tunis. Husayn's choice not to own land or buildings in Florence, and the resulting limitation of his local rights, made his presence in Tuscany and European lands more precarious and less readily visible.[96]

In this he was similar to other Muslims residing in Europe. During the second half of the nineteenth century, Jewish and Muslim North Africans who settled in Europe were able to buy land and buildings, but few exiled Tunisians did so. Here a major dissymmetry comes to the fore: southern lands exerted a strong pull on Europeans, but northern lands did not attract Maghrebi investors.

THREE

A World of "Affairs"

Litigation as a Tool for Negotiation

THE FIRST chapter shed light on Husayn's life and afterlife, including his extensive involvement in the Ottoman reform project; the second chapter showed the significance of owning land for Husayn and other Muslim subjects, both before and after the French conquest of Tunisia. This third chapter sets Husayn's life and legacy even more broadly against the backdrop of the second half of the nineteenth century, when conflicts over Tunisian dignitaries' estates resulted in major litigation involving both European and Ottoman states. In this chapter, I will trace Husayn's political actions through his involvement in significant precolonial and colonial-era court battles over the assets of two former dignitaries from Ottoman Tunis: Mahmud Ibn 'Ayyad (figure 3.1) and Nessim Scemama.

In the mid-1850s, Husayn accompanied his patron Khayr al-Din to Paris to file a complaint against Mahmud Ibn 'Ayyad, a former Tunisian tax collector who had left Tunis for Paris in 1852 after having ruined the tax farms in his charge.[1] Husayn also acted on behalf of the Tunisian government against Nessim Scemama, a Tunisian Jew who replaced Ibn 'Ayyad as the main tax collector and Tunisian state treasurer before embezzling public funds and leaving Tunis for Europe in 1864. When Scemama passed away in Livorno in 1873, Husayn contested the validity of Scemama's will in the Italian courts in order to retrieve some of the public funds Scemama had embezzled.

In this chapter, I argue that we cannot fully understand the beginning of French colonization in Tunisia and Husayn's agency during the colonial

FIGURE 3.1 Mahmud Ibn 'Ayyad, French National Library, Paris. *Source:* Jean-Baptiste Adolphe Lafosse (1810?–1879), Le Général Sidi Mahmoud Benayet, ministre du commerce et de l'agriculture de S.A. le bey de Tunis, after a painting by Charles Gleyre, BnF, département des estampes et de la photographie.

period if we do not consider these two major cases. Indeed, they dragged on for decades: the battle over the Scemama estate was only settled in the 1890s, that of Ibn 'Ayyad two decades later. I will show that these conflicts strengthened the legal and financial relationships between France, Italy, and Tunisia more than three decades before the French conquest of Tunisia in 1881. I will also contend that these legal battles generated an increasing number of other disputes that empowered Ottoman dignitaries and Tunisian subjects like Husayn to act against European powers and banks, including during the colonial period. Husayn's contemporaries were aware of the prejudicial climate and of the increasing number of legal disputes in the wake of the Ibn 'Ayyad and Scemama cases. One of Husayn's former intimates in Tuscany, the lawyer David Santillana, who worked for the Tunisian government, deplored the "very confused, very tangled" situation in 1884, lamenting that "one constantly sees [trials] in the courts of Paris and Livorno . . . against former accountants and ministers."[2]

Articulating the long-lasting effects of these cases is crucial for a more general reason. Historians have mainly scrutinized colonial legal disputes to argue that they helped structure the colonial legal systems or, on the contrary, to defend the idea that litigation ended up tearing the colonial regimes apart.[3] While taking these hypotheses into account, I do not look at the Husayn, Scemama, and Ibn 'Ayyad affairs as illustrating either the fragmentation or the structuration of a colonial order. Rather, I view them as episodes revealing broader historical changes within Mediterranean societies.[4] In the first section of the chapter, I present the roots and development of the Ibn 'Ayyad and Scemama legal cases, together with the numerous disputes they prompted. I argue that these disputes were at the same time the reason for and the result of increasing financial and legal entanglements between European and Tunisian societies that historians of modern North Africa have strangely neglected.

In the second section, I explore the extent to which this litigation provided North Africans like Husayn with tools for action and negotiation. European and Tunisian stakeholders branded these cases "affairs"—*affaire* in French, *affare* in Italian, and *nawāzil* (sing. *nāzila*) or *qaḍiyya* in Arabic—a widely used moral category that referenced notions of public scandal and financial corruption. For Europeans, these "affairs" were evidence that the Tunisian government was unable to properly administer its own state; the French in particular used these "affairs" to justify their interventions in Tunisia, up to the 1881 colonization, as attempts to reorder the ostensibly corrupt Tunisian state. Husayn also referred to this moral category during missions in Europe, but for a very different purpose: to defend Tunisian interests and the necessity of retrieving the public funds that Ibn 'Ayyad and Scemama had invested in Europe. For many individuals, including Husayn, the Ibn 'Ayyad and Scemama legal conflicts also provided some room to maneuver, and even financial and legal resources during the French colonization of Tunisia. For instance, rather than helping settle the disputes over the Scemama estate, Husayn deliberately prolonged them in order to exert his agency from 1881 to 1887, when France already ruled Tunisia.

But during the colonial period, historical actors like Husayn and French colonial representatives did not fully control the multiplying legal disputes that arose from the Ibn 'Ayyad or Scemma affairs, as we shall see in the third section of this chapter. On the one hand, intermediaries—mainly interpreters and brokers of Jewish origin working for Husayn in Tuscany—used the disputes surrounding the "affairs" to turn against the Tunisian state and negotiate new social positions vis-à-vis the French colonizers. On the other hand, these cases prompted a deluge of written texts that swamped the courts, European consulates, and newspaper columns, overwhelming the various players in the affairs. The profusion of primary sources was nothing new in Muslim and European societies, which were very litigious.[5] Nonetheless, as we shall see, the number of papers, written proofs, and legal arguments increased even more in the second half of the nineteenth century for at least two major reasons: first, because of the growing use of printed material in the Ottoman Empire; and second, because of the ever-increasing value placed on written evidence in the various legal systems of Europe and North Africa. This profusion of writing by the second half of the nineteenth century contradicts the commonplace idea in colonial historiography that there are few or even no Tunisian sources in Arabic.[6] This profusion of writing, in turn, also generated other legal disputes.

The Ibn 'Ayyad and Scemama "Affairs": Two Legal Cases Entangling Tunisia and Europe

Mahmud Ibn 'Ayyad and Nessim Scemama, the two figures at the center of the major legal disputes relating to the misappropriation of Tunisian public funds from the mid-nineteenth century onward, were, successively, in charge of the Tunisian treasury during the Ottoman reform period, from the 1830s to the 1860s. Mahmud Ibn 'Ayyad was the trusted servant of the Tunisian sovereign Ahmad Bey, who reigned from 1837 to 1855. As such, he was the director of finance, a major tax collector, and the administrator of tobacco, leather, and salt. He was also in charge of coin-making and was "general supplier" to the Tunisian army established in the wake of reforms in the early 1830s.[7] Ibn 'Ayyad was the first Tunisian dignitary to misappropriate public funds and divert them to Europe, fleeing to France with funds from the Tunisian bank that he had created in 1847.[8] The Tunisian authorities seized some of his assets held in the Ottoman Empire after his departure for Paris in 1852 and his naturalization as a French citizen.[9]

The second figure, Nessim Scemama, was indebted to Ibn 'Ayyad. He held similar positions to those of his former superior after Ibn 'Ayyad left for France. Scemama controlled nineteen tax farms between 1857 and 1864, and supervised government expenditure.[10] Scemama used several techniques to amass his fortune, to the detriment of the Tunisian State.[11] For instance, he delayed paying government securities for several months. The holder of a security, often "needy and in a hurry, would therefore decide to sacrifice a quarter or third of the sum that was owed him," and sometimes more. The security holder would sell his security to one of the brokers working for Scemama, who would then pocket the difference.[12] The need to carry out the Ottoman state reforms from the 1830s to the 1860s led to a vast increase in financial missions. The urgent need to reform the Tunisian administration together with the limited supervision of state funds meant that tax collectors and state treasurers such as Ibn 'Ayyad and Scemama could exploit the situation and work with Europeans to embezzle public funds.

Although originating from the same context, the disputes over the estates of Ibn 'Ayyad and Scemama did not take exactly the same form. In the first case, it was Ibn 'Ayyad and then his heirs who sued the Tunisian government, claiming that the Tunisian state had to return the sums seized from the Ibn 'Ayyad family. In the early 1850s in Paris, where Ibn 'Ayyad sought refuge, the French Ministry for Foreign Affairs set up a dispute committee acknowledging that Ibn 'Ayyad was a creditor of the Tunisian government. Ibn 'Ayyad and, after his death in Istanbul in 1880, members of his family brought claims against the Tunisian government, first before the civil court in Paris and then in Tunis after the French occupation in 1881. The Ibn 'Ayyads argued that they were creditors of the Tunisian government for increasingly large sums of money: more than 37 million French francs in 1886, and 237 million French francs in 1903.[13]

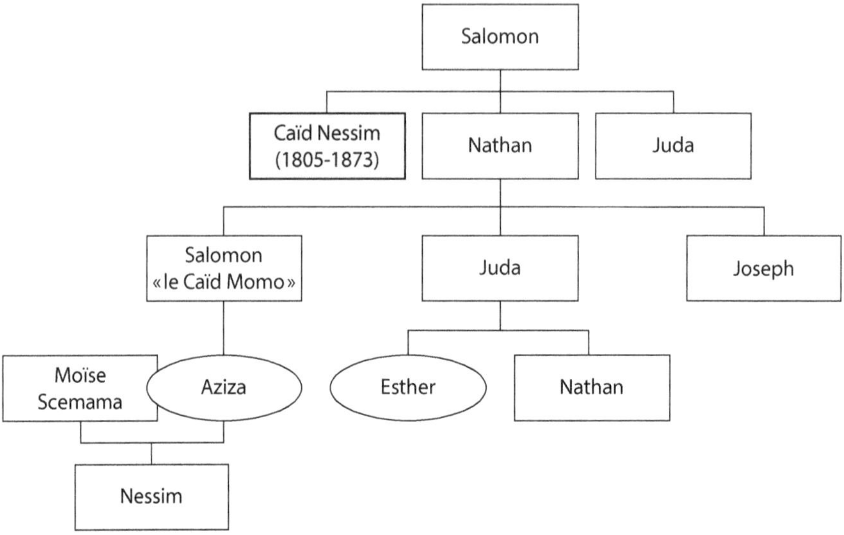

FIGURE 3.2 The Scemama family

In the case of Nessim Scemama, the disputes over his estate, which lasted more than two decades, were primarily related to his succession. Scemama did not have any children. The instructions in his will, revealed at his death in Livorno in 1873, stated that half of his estate should go to Aziza, the daughter of his nephew Salomon, whom he had brought up, and to Aziza's son, who was also called Nessim. A quarter was to go to his nephew Joseph, and the last quarter to his great-nephew (figure 3.2).[14]

Husayn, representing the Tunisian state, contested the legitimacy of Scemama's will on two counts. First, according to Husayn, the estate had stemmed from an immense fraud "against the Treasury."[15] Second, Nessim had not written the will in accordance with the rules imposed by halacha, the body of Jewish laws derived from the Torah. On the opposing side, Scemama's niece Aziza and one of his two wives sought the backing of European powers, especially Italy.

The Ibn 'Ayyad and Scemama legal cases were critical. They were related to other disputes between the Tunisian state and former administrators during the 1870s and 1880s, including during the early years of colonization.[16] In 1872, Scemama's nephew, Salomon, treasurer to the Tunisian government, was placed under supervision, and his trips to Italy were overseen.[17] The following year, he fled to the Greek island of Corfu. In total, Salomon misappropriated "over 10 million Tunisian piasters between 1864 and 1866."[18] About fifteen years later, General Tax Inspector Caid Elyahu Scemama was suspected of having embezzled 2,700,000 Tunisian piasters; he was stripped of his office and title.[19] Ibn 'Ayyad's nephew, Hamida, claimed in late 1883 that the Tunisian government owed him four million piasters; he was also party to a series of real estate disputes in Tunis. He placed himself under the legal protection of the British government and used this protection to press claims before the French tribunal in Tunis and the French Court of Appeals in colonial Algiers.[20]

There was nothing new about using European legal systems to settle affairs concerning Tunisian claimants. Muslims and Jews from North Africa had filed claims and been involved in disputes in Livorno, London, and other major trading centers in Europe throughout the early modern period.[21] But the Ibn 'Ayyad and Scemama cases were of a completely different financial and legal scale. Considerable sums of money were at stake.[22] National and transnational legal hierarchies were brought into play on both shores of the Mediterranean—in Tunisia, Algeria, France, Italy, and Istanbul—over the course of proceedings lasting more than half a century in the Ibn 'Ayyad case. Not only do the Ibn 'Ayyad and Scemama affairs reveal intense interaction between the two shores of the Mediterranean, they also contributed to the entanglement of North African and European societies, financial networks, and legal systems, both during the early years of colonization and in the long run as well.[23]

France and Italy offered Ibn 'Ayyad and Scemama legal protection and even European citizenship, allowing the two men to flee from Tunisia with their capital: Ibn 'Ayyad chose to become French, and Scemama considered becoming Italian. Both men transferred colossal sums to Europe and thus fueled the banking institutions that were developing more sophisticated credit techniques in the second half of the nineteenth century.[24] By plundering the government they were meant to serve, these men not only weakened the Tunisian state, they also bound Tunisia more tightly to Western Europe. The Ibn 'Ayyad and Scemama cases transformed European and North African societies in yet another way: they helped fashion moral discourse and served as the basis for legal and financial action by Europeans and Tunisians on both shores of the Mediterranean.

The "Affairs" as a Tool for Action and Negotiation

"Affairs" as a Moral Concept

Europeans and Tunisians both called the Ibn 'Ayyad and Scemama cases "affairs." This notion implied a public scandal as well as instances of financial corruption. Like many other European "affairs" at that time (although unlike the Dreyfus Affair that shocked and divided France in the late nineteenth and early twentieth centuries), the Ibn 'Ayyad and Scemama affairs neither profoundly subverted European public opinion nor revealed a new moral vision.[25] Yet the categorization of Ibn 'Ayyad's and Scemama's legal disputes as affairs "emphasized the gap between a normative ideal" and a corrupt world.[26]

It is true that European, particularly French, observers drew on this moral repertoire. European and French diplomats, bankers, and journalists brought the notion of "affair" with them to Algeria and Tunisia and used it to denounce dysfunctional local governments, financial disorder, administrative incapacity, and even a lack of civic-mindedness.[27] As we saw in the second chapter, the Tunisian state ran up public debt in the 1860s and was declared bankrupt in 1867. Publicizing the Ibn 'Ayyad and Scemama affairs was a way of assigning most of the responsibility to the Tunisian government, which was seen as unable to administer its public finances.[28] This discourse intersected with Orientalist ideas about the corruption or decline of the Muslim world and the Ottoman Empire.

Employing such moralistic language was a way of sweeping under the rug the fact that elites on *both* sides of the Mediterranean were responsible for Tunisian bankruptcy. Stressing the responsibility of Tunisian elites allowed the French to hide the involvement of financiers at Paris-based Erlanger Bank in the 1867 Tunisian bankruptcy and in the management of the Scemama estate. Tunisian officials first approached Erlanger Bank in 1863 for a public loan with extortionist terms. Ten years later, when Scemama died, Erlanger Bank bought up parts of his estate from his heirs.[29] Above all, Erlanger Bank managers wanted to sideline Husayn as the Tunisian government's representative in the Scemama case. In 1881, Erlanger bankers succeeded in convincing the French colonizers to replace Husayn with Giaccomo Guttierez, a financier more amenable to the bank's interests.[30] Erlanger Bank's preeminence in Tunisian affairs deeply concerned Husayn, who stated in 1884 that "the Tunisian authorities no longer own a single camel, neither male nor female. Erlanger is the master of this affair. It bends the government to its desires (*yatafaḍal 'alā al-ḥukūma in shā'*)."[31]

But it would be wrong to think that European and French colonial actors were the only ones to draw on the moral implications of the notion of "affair." Husayn drew on the Arabic equivalent of *nāzila* (a specific case).[32] He was clearly aware that Ibn 'Ayyad and Scemama had irrevocably harmed the Tunisian state budget. Husayn continually refused to meet Ibn 'Ayyad when on missions in Europe, while Ibn 'Ayyad and his family maintained connections with other Tunisian statesmen outside of European courts.[33] In the 1870s, Husayn developed the same attitude toward Scemama. In legal memos, he argued that Scemama had committed an immense fraud "against the Treasury."[34] Finally, in addition to shaping the moral categories used by both Tunisians and Europeans, the Ibn 'Ayyad and Scemama affairs provided Husayn with a way to defend his country's interests in an asymmetric political world at a time when the Tunisian government was becoming powerless in the face of European legal and administrative systems.

How Husayn Complicated the Scemama Case

From 1873 to 1881, Husayn was entrusted with defending the rights of the Tunisian government in the Scemama estate trial. He acted on two fronts. Initially, Husayn adopted a legalistic approach, hoping to act via the Italian judicial system.[35] In the Tuscan courts, he argued that Scemama's will was illegal and that the deceased was of Tunisian nationality, in a bid to legitimize the Tunisian government's claim to Scemama's estate. But the rulings against his arguments led him to launch a series of appeals over more than seven years,[36] and Husayn eventually lost faith in the Italian system. In defense briefs, he stated that the Italian courts were largely unable to take into account the legal customs and lines of argument prevalent in Islamic countries.[37] Accordingly, Husayn also operated via quasi-official channels. Instead of public legal debate, he used secret procedures, clandestine negotiations, and false documents.[38] From the late 1870s onward, Husayn used each of these secret financial arrangements, sometimes quite deliberately, to prompt new legal disputes, a tactic that would allow him to continue acting in these affairs during the colonial period.

One of the ways Husayn sought to recover the Scemama estate was by obtaining financial securities that had not figured in the estate inventory. For instance, Husayn and his assistants found a Fr.500,000 note that Scemama had signed just before leaving Tunis in 1864. Hay Sabbagh, formerly one of Scemama's servants, notified the Tunisian government that he held the note, which was blank, meaning it had no bearer's name.[39] Acting with his government's consent, Husayn officially seized the note for philanthropic purposes, arguing that half of the sum to be recovered on the note would be used to found an Islamic hospital.[40] But in order for the Fr.250,000 to revert to the state, Husayn needed a different bearer than Sabbagh, who was not permitted to testify in the Muslim courts in Tunis because of his Jewish faith. Instead, Husayn sought a trustworthy Muslim who could provide oral testimony asserting to the veracity of the document and who would agree to have his name on the note without trying to appropriate the sum for himself. In 1878, Husayn made out the note to one of his friends, Mustafa al-Murali, a colonel living in Tunis.[41] Al-Murali was reminded that "he had no rights to the security bearing his name" and that it belonged solely to Husayn,[42] who agreed to pay all the legal costs necessary to recover the security and to defend al-Murali from any legal attacks.[43] After Hay Sabbagh died in 1879, Husayn drew up a new secret agreement with Sabbagh's two sons and heirs, backdated to May 2, 1878.[44] In this agreement, Husayn acknowledged (albeit unofficially) that he and the two sons, Elyahu and Joseph, jointly and equally owned the security.[45]

In 1884, even though he had ceased to represent the Tunisian government in this affair three years before, Husayn used the note in proceedings against Scemama's heirs. The Livorno court found in favor of al-Murali and sentenced Scemama's estate to pay back the Fr.500,000 at 5 percent interest.[46] Victory was within Husayn's grasp, but at the last minute, his right-hand man, Léon Elmilik, revealed that the evidence on which the legal proceedings were based was partially false. Elmilik turned against his employer, claiming that Husayn owed him a significant portion of his salary and other funds. Elmilik obtained a declaration in Judeo-Arabic from Sabbagh's heirs admitting that the note had been blank and that the Fr.500,000 belonged solely to their father. The two Sabbagh sons agreed to pay Elmilik 12.5 percent of any sums that came into their possession.[47]

In 1887, after a further series of twists and turns, the opposing parties were exhausted. The Tunisian state and Erlanger Bank, which had granted a 12-perent-interest loan to the state and bought up parts of Scemama's estate, agreed to negotiate. Al-Murali and Elyahu Sabbagh negotiated a settlement of Fr.325,000 for the Fr.500,000 bill.[48] Al-Murali and Husayn received half of this sum, or Fr.162,500. After legal costs, only Fr.52,828 remained.[49] This does not appear to have been used for philanthropy, as Husayn had promised. Nevertheless, Husayn had managed to complicate any settlement of the Scemama affair. By deliberately provoking legal disputes, he prevented both Scemama's heirs and the Tunisian state—now colonized by the French—from laying hands on the funds.

Husayn and his agents also pursued a second strategy to recover some of the money misappropriated by Scemama. They sought to sell the Tunisian government's claims on the Scemama estate to a group of French investors. Any rights sold would help bail out the Tunisian state, which had been declared bankrupt. In 1879, Husayn wrote that he was unable to sell these claims to a Marseille-based company, which had "doubts about the exactitude" of the Tunisian government's accounts.[50] Instead, Husayn went to Rome to sell the claims to another "group of capitalists," of whom "one is French, another English, the third German, the fourth Swiss, and the two others are Italians."[51]

Husayn argued that it was necessary to act in secrecy, for if "our opponents notice" and "know the sum, they may intervene in the affair and seize the transfer."[52] On November 7, 1879, without the backing of the Tunisian government, Husayn signed a contract with a certain Mayer, who was acting on behalf of Count Joseph Telfener, a businessman, engineer, and worldwide railway developer.[53] In 1881, Husayn's replacement in the Scemama affair, Giaccomo Guttierez, reached a compromise compensating the new investors: the Tunisian government was to pay Fr.800,000 to Count Telfener as soon as the Scemama estate was settled.[54]

Husayn's real role in this transaction was discovered only after he had passed away, when his estate was liquidated and Angiolina Bertucci sold a declaration to Erlanger Bank.[55] In this declaration, dated August 1, 1881, Count Telfener acknowledged that he transferred "all rights in the affair" to General Husayn.[56] Husayn thus deliberately complicated matters when it came to the management of the Scemama estate, as he had in the case of the Fr.500,000 note. He brought in dummy owners, accomplices (al-Murali and Santoni), and a seasoned businessman (Telfener).[57]

Husayn initiated legal disputes and proceedings to defend his interests and those of the Tunisian state at a time when Ottoman dignitaries from Tunis were no longer able to fight European and imperial administrative hierarchies on equal footing. Husayn had no official mandate as of 1881. He was unable (or unwilling) to rely on European legal systems, whose principles he viewed as unfair. As a sidelined agent during the colonial period, though one still driven by Ottoman and Islamic loyalties, he counted on ruses and secrecy to promote the interests of a financially ruined Tunisian state.[58] Yet neither Husayn nor the French colonial representatives fully controlled the legal disputes or their outcomes. Indeed, these cases became even more complicated by the late 1870s, when North African Jews, who had acted as intermediaries between Tuscany and Tunis since the sixteenth century, began to use legal disputes to promote their own interests and to negotiate new positions within the French colonial system.

The Litigations That Overwhelmed Tunisian and European Societies

How Translators and Brokers Intensified Legal Disputes

Many historical studies have examined the role of Jews of Livorno as intermediaries between Europe and the Maghreb during the precolonial period.[59] These men used their command of Arabic, Judeo-Arabic, and European languages to represent the interests of Muslims who either were unwilling to travel to Christian lands or had been sidelined by European rivals.[60] Historians of early modern Tunisia tend to group them together under the term *Qrāna* (Livorno, in Arabic) as opposed to Jews from Tunis, referred to as *Twānsa* (Tunisians), each community having its own specific organization and religious hierarchy. When these communities are studied in the context of Livorno, the Jews from the Maghreb are distinguished from European Jews in Tuscany.[61] However, historians have paid far less attention to the dependent relations these North African and Italian Jews had with their Muslim patrons, or to the role they played in the legal disputes that linked the two shores of the Mediterranean in the second half of the nineteenth century.

Prior to the colonial period, Husayn did not depend on Jews from Livorno to carry out his mission regarding the Scemama estate. He recruited North

African Jews and Muslim employees who assisted him in his missions in Tuscany, and over whom he had a certain amount of control, in Tunis and Algiers. In Florence, Husayn collaborated with attorney David Santillana, son of Moïse Santillana, a former interpreter to the governors of Tunis who had worked for the British consulate.[62] David Santillana was also hired as an auxiliary interpreter by the Tunisian Department of Foreign Affairs.[63] Léon Elmilik became associated with Husayn's mission in a similar fashion. Elmilik originated from neighboring Algeria and, like all Jews in the northern part of the colony since the 1870 Crémieux decree, had French citizenship.[64] He came recommended to Husayn in 1877 by Hay Sabbagh, the former servant of Scemama and Prime Minister Khaznadar.[65]

It was also in Tunis that Husayn recruited Salomon Rosa, the son of another of Scemama's former servants. Rosa had built his career in the shadow of his protectors. He had followed Scemama to Paris in 1864—Rosa's father had begged Scemama to take his son "even as a servant" who would "ask no salary of him, and would be happy with just his room and board."[66] When Scemama fired Rosa in 1867, accusing him of betrayal, Rosa sought the protection of another Tunisian dignitary, General Muhammad al-Bakkush, who had come to negotiate the Tunisian debt.[67] Bakkush took pity on "this child far from home who had been abandoned in poverty" in Paris and "paid for his return to Tunis." Rosa eked out a living working for the Tunisian Department of Foreign Affairs before going to Florence as an interpreter with Giacomo di Castelnuovo, former physician to the Tunisian governors and the new Tunisian consul in Tuscany. Rosa was sent back to Tunis, but returned eventually to Tuscany in 1874. He entered the service of Husayn, to be involved in litigation against Rosa's former employers: Castelnuovo and then Scemama. He became part of Husayn's inner circle, even eating at his table. Husayn financed the studies of two of Rosa's brothers and had his family housed "in a magnificent mansion" in Tunis.[68]

Husayn saw these employees as his servants whom he remunerated on behalf of the Tunisian state.[69] Their specific knowledge would allow him to infiltrate Livorno's Jewish communities and gain access to their writings. Their principal mission was to read and decipher various Judeo-Arabic documents pertaining to the Scemama case.[70] Rosa, Elmilik, and then Santillana were tasked with finding written evidence to present before the Italian courts.[71] They had to enter into contact with brokers in town to set up financial arrangements and buy parts of the Scemama estate. Initially acting as translators, document collectors, and brokers for their employer, Rosa, Elmilik, and Santillana went on to manage Husayn's accounts and personal affairs. Elmilik helped Husayn obtain disputed payments in Tunis and negotiated loans.[72] In 1882, it was Elmilik who drew up a rental agreement (*'aqd*) for the French army to take quarters in Husayn's palace, located at one of the main entrances to the medina of Tunis.[73]

By the late 1870s, as French influence grew in Tunis, Husayn's patronage of Rosa and Elmilik and his intimate ties with them no longer sufficed to control their actions. After serving Husayn from 1878 to 1883, Rosa and then Elmilik turned against their employer. Husayn's intermediaries realized that events were at a tipping point and that France was acquiring a dominant role. They complicated the legal situation even more by initiating new disputes in the Scemama affair, against Husayn. After Rosa left Husayn's service, his former employer accused him of purloining account registers.[74] Husayn also suspected Rosa of having spied on him for two years in exchange for monthly payments of Fr.5,000 from Erlanger Bank. Rosa started working for France, becoming an interpreter at the French consulate in Livorno.[75] Elmilik, Rosa's former father-in-law, stole thirteen sets of documents from Husayn in 1885, including the registers for *inzal* tenancies at La Goulette.[76] He sold them to Erlanger Bank for Fr.6,000, together with claims to the Scemama estate and the Fr.500,000 note in al-Murali's name.[77] Elmilik accused Husayn of not having paid him enough, claiming Fr.175,000 in back wages. Husayn, in turn, called for Elimilik to present his accounts. A vigorous legal tussle ensued between the two men in the courts of Tunis and Livorno, even reaching the Court of Appeals in Florence, despite attempts at conciliation, including an initiative by Elmilik's daughter.[78] The two parties continued to sue each other in Italy and Tunis until Husayn passed away.[79]

Involvement in the Scemama affair enabled these brokers and interpreters to expand their social horizons and business networks. They even paid their professional expenses out of pocket and would loan money to their employer, Husayn.[80] But intermediaries such as Rosa and Elmilik were no longer useful to the Tunisian state in the colonial context, creating more problems than they were worth when the Tunisian administration fell under French control. Furthermore, French colonial authorities regarded them as amoral. In a short, anonymous note sent to the French consul in Livorno, Elmilik was even accused of having "sold his daughter" to Husayn and "profit[ing] from this dishonorable position to administer the fortune of his daughter's lover for his own advantage."[81] Rather than serving their government, each sought to press claims against the Tunisian state, adding to the era's many legal disputes. French officials had to negotiate with these intermediaries to get rid of them and their claims.

In addition to the unexpected changes in their mediators' allegiances, the men and women involved in the Ibn 'Ayyad, Scemama, and Husayn legal cases had to cope with yet another complication: the dizzyingly increasing volume of written documents produced in the midst of such legal disputes: originals, copies, genuine-but-altered documents (such as the Fr.500,000 note), forgeries, and other writings that were stolen or sold to Erlanger Bank triggered further litigation and "affairs" both before and during the colonial period.

An Overwhelming Profusion of Texts

In late 1879, as he was dealing with the Scemama case, Husayn was drowning in a sea of documents. After six years in Tuscany, Husayn and his employees had inventoried fifty thousand documents related to the Scemama estate, of which more than a quarter still needed to be stamped with two different seals.[82] The sheer volume of written documents in this affair was not unusual. Contrary to the idea of an archival desert—a commonplace in historical scholarship about colonial North Africa—North African societies were as litigious as their European counterparts, and from the fifteenth century onward, writing, "in the form of countless notarial deeds, epistolary correspondence, and community regulations," spread across the various parts of the Maghreb, even to territories as remote as southern Morocco.[83] Under Muslim law, oral testimony (*shahāda*) had always been fundamental, but writing was a crucial support to oral testimony.[84] Written documents had to be protected, because they were vulnerable.[85] They could be torn. A testimony or even a word could be difficult to read and understand, or might even have been altered or falsified while being transcribed.[86] Written documents became a "vital personal effect" that elite and common people alike protected and conserved, folded or rolled up in a safe place.[87]

In the late nineteenth century, Husayn, his contemporaries, and the men governing Tunis faced similar issues in authenticating and handling testimony and evidence. In addition, they had to deal with more than just the Islamic legal system and the central role of oral testimony—they also had to make their case in European courts and therefore needed knowledge of European ideas regarding evidence and oral testimony. Even in precolonial Tunisia, subjects and dignitaries were accustomed to the various understandings of evidence and testimony within a legal system that had already become plural: Islamic sharia and Jewish rabbinical courts coexisted with state or civil courts implemented during the Ottoman reform era and with European consular courts that were involved in cases related to European communities. After taking over Tunisia, the French reshaped this legal pluralism according to their own needs: in addition to the Islamic and rabbinical courts, in 1883 they established French courts, which had jurisdiction over French citizens and "nationals from foreign countries who renounced" their specific rights and legal privileges.[88] This legal pluralism prompted increased production of legal documents, such as notarial and property deeds, and divergent approaches to their meaning and importance.[89] Across the Ottoman Empire, state administration and judges increasingly recognized written evidence in the early modern and modern periods, particularly following the Ottoman civil code promulgated in 1877.[90] This code established court registers (*sijillāt*) as providing probative value.[91] Confronted with all these legal reforms and systems, men like Husayn had to be conversant with both new and old norms and techniques from European and Muslim law in order to stay on top of an ever-increasing body of written evidence.

In addition to the changes within, and the intersections between, the various legal systems, there was yet another reason why writing became ever more entwined with the spoken word when it came to proof and testimony. Throughout the nineteenth century, new government techniques transformed administrative procedures for identifying potential witnesses and parties to a claim.[92] The states were strengthening their capacity to identify individuals, follow their movements, and control them. Bureaucrats began to issue administrative documents, such as identity cards, passports, and transit passes. In Islamic law, witnesses (*'udūl*; sing. *'adl*) who could "bear truth" were chosen from among free Muslims judged virtuous and competent by their communities,[93] but as shown by Brinkley Messick, these criteria became less relevant when states were able to identify their citizens and check their records and reputation from a distance.[94] In such modern states, with their accompanying legal apparatus, the fear of the unknown witness and doubts about a witness's reputation become less of a concern. Almost any individual could be identified and the person's testimony transcribed.

The new force of writing and of written evidence was related more broadly to new forms of education, new relationships between oral and written language, and new technology such as Arabic type for the printing press, and the telegraph. As we have seen, in the Tunisian governors' palaces in the 1830s and 1840s, Husayn had experienced the privileged position that Muslim scholars accorded to oral transmission and recitation within a circle of students (*ḥalqa*). At the Military School of Bardo, he was trained in reading and writing with texts and books that were either in manuscript or printed.[95] Later on, in Tuscany, Husayn and his employees used handwritten documents as much as printed ones. In addition to their vast correspondence and travelogues, they published memoirs and legal pamphlets detailing their cases and judicial strategy. They financed newspapers and responded in their pages to the criticisms and defamatory observations leveled against them by their European rivals. Printing techniques using Arabic characters and lithography spread to the Maghreb from the island of Malta (under British rule) and from Egypt, which had a printing house in the Cairene neighborhood of Bulaq as early as 1822. This spread of printing techniques across the Ottoman Empire led, during the second half of the nineteenth century, to handwritten administrative documents' slowly being replaced by preprinted forms that administrators and individuals could manipulate (or falsify).[96] At the same time, Husayn and his contemporaries intensified this production of documents and accelerated the dissemination of information using new techniques such as telegraphy.[97]

The new importance accorded to written testimony, the transformations of plural legal systems, the new administrative methods to identify individuals, and the spread of new techniques such as the printing press all fueled a tidal wave of paper that consequently gave rise to new claims. In the Scemama affair, and later in the conflicts over Husayn's legacy, individuals resorting to the various legal systems in Tunis, Paris, and Tuscany requested authentication of translations, duplicates, and copies for their many claims.[98] As a result, bureaucrats were unable to oversee the increasing number of written documents. In Tunisia, neither local nor French officials were able to find enough interpreters to read and translate every document.[99] In 1880, one year before being stripped of his functions as a government representative,

Husayn complained that he was alone in Florence "without anyone to help me with these documents in Arabic."[100] Government representatives such as Husayn and his lawyer Santillana took much of the reading and writing upon themselves. When Husayn passed away, Santillana worked with another interpreter from Tunis to carry out the "operations to go through and analyze" the papers relating to Husayn's estate.[101]

Besides insufficient numbers of skilled translators, the Tunisian and French administrations had to deal with the scattering of documents across several territories. Official documents and private papers related to the Ibn 'Ayyad and Scemama affairs and then to Husayn's estate were held in various locations, including government offices and private residences. French and Ottoman representatives hoping to collect papers relating to the Scemama and Ibn 'Ayyad affairs from Husayn's estate after 1887 mention disorderly documents and files. The Ottoman consul Vasilaki Musurus Ghikis said he "found the two houses" Husayn rented in Florence "in the greatest mess," with papers and letters scattered all over the place.[102] In fact, even when Husayn was representing the Tunisian state in the Scemama affair, he could barely locate all the documents related to the case. In the heat of legal disputes, pieces of evidence were simply not returned after their use in court. They remained in the hands of Husayn's private secretaries, who sold some of them to the highest bidders—including to the Parisian Erlanger Bank.

For Husayn and his entourage, the distinction between official and private papers was not obvious. When asked to account for his management of the Scemama case in 1881, Husayn could not make a clear distinction between the papers belonging to himself and those belonging to the Tunisian state.[103] He was also reluctant to hand over documents to the Tunisian government that had disgraced him and from which he had hidden negotiations, the forging of evidence, and other surreptitious maneuvers. In 1876, Husayn sued the Department of Justice in the Italian Ministry of Foreign Affairs, which claimed that "the state is supposed to be the sole owner of official pieces of evidence, even though these may, due to a variety of circumstances, be in the hands of a mere individual."[104] In such circumstances, even the rights of various parties to hold documents became a source of legal dispute between the Tunisian state, its representatives, and its former servants.

At first sight, the legal disputes at the core of this chapter do not seem remarkable. Unlike the popular uprisings that burst out in modern North

Africa, they did not prompt rallies and protests. However, these legal conflicts were instrumental in revealing other power dynamics, particularly forms of underground action rooted in alternative uses of moral discourse. Because they involved many spheres of jurisdiction and massive public funds, the Ibn 'Ayyad and Scemama affairs strengthened the web of legal and financial connections between France, Italy, and Tunisia beginning in the mid-nineteenth century, long before European colonization in the Maghreb, and extending more than three decades after the French conquest of Tunisia. Additionally, because stakeholders in these affairs managed to complicate and even worsen these legal disputes, the Ibn 'Ayyad and Scemama cases allowed dignitaries and individuals such as Husayn and Léon Elmilik room to maneuver during the colonial period, to the point that they could negotiate new social and financial positions with the French authorities. Finally, the Ibn 'Ayyad and Scemma affairs fueled other disputes that no one—neither North Africans nor Europeans—could fully control. One of the unexpected consequences was the increasing amount of paper—administrative documentation and written evidence—that prompted new claims and disagreements.

The Ibn 'Ayyad and Scemama affairs were related to other major transformations that I have examined in the two previous chapters and that had significant effects on Husayn's life and afterlife—namely, the Ottoman reform program that, since the 1830s, had led, on the one hand, to a major financial crisis in the late 1870s and, on the other hand, to the exile of North African elites in Europe over the course of the second half of the nineteenth century. These major historical changes, along with the legacy of Ottoman rule, will resurface in the next three chapters as we will deal with yet another case, the Husayn affair, which encompassed the many conflicts waged over Husayn's legacy during the French colonial era in Tunisia from 1887 to the late 1920s.

FOUR

The Diplomatic Conflicts Over Husayn's Estate

Ottoman and Italian Interventions

JUNE 27, 1887. At 3:30 p.m., four hours after Husayn passed away, in the sticky afternoon heat, the Ottoman consul in Florence, Basile Musurus Ghikis, had no other choice than to go swiftly up to the hills surrounding the capital of Tuscany and perform his consular duties: as quickly as possible, he placed seals on the doors of Husayn's apartments in the Sicard villa as well as on the body of the deceased,[1] then seized the two horses that had belonged to Husayn.[2] The consul had acted fast. His French counterpart in Florence, theoretically in charge of Tunisian citizens since the French colonization of Tunisia, took nearly three days to place his own seals on the outside doors of the Sicard villa, preventing its owner, Mr. Sicard, from gaining access to his own apartments.[3]

Musurus Ghikis was not the most zealous diplomat in Florence, however. In letter after letter following his arrival, the Ottoman consul expressed his doubts and sadness about his appointment to his cousin Étienne and to his uncle and protector, Constantin Musurus, the long-time ambassador to London for the central government of the Ottoman Empire, also known as the Sublime Porte. Musurus Ghikis did not understand why the Ottoman authorities had sent him to a demeaning, second-rank post in Tuscany.[4] The consul's ordeal was not yet over. In the coming days, after placing the seals on Husayn's body, Musurus Ghikis had to hand over Husayn's mortal remains to the Ottoman consul in Livorno for burial in the Muslim cemetery there. Musurus Ghikis also had to compete with his

French counterpart, Consul de Laigue, who held him in contempt, finding him to be "hopeless," "debt-ridden," and so sick as to be unable to leave his residence.[5] For three and a half months, from late June to September 1887, the Ottoman consul and his French counterpart fought over virtually every aspect of the Husayn estate.[6] They heatedly debated the Tunisian or Ottoman nationality of this former Circassian slave who had been educated in Tunis and exiled in Tuscany. Musurus Ghikis was even involved in one final, awkward act: digging up Husayn's remains from the Muslim cemetery in Livorno to transport them to Istanbul for reburial in an Ottoman mausoleum.

This chapter focuses on the many diplomatic conflicts over Husayn's legacy, involving four main players: France, Tunisia, the Ottoman Empire, and, to a certain extent, Husayn's host country, Italy. In this chapter, I show that French colonial authorities were not the only foreign officials interested in Tunisian matters; representatives of other powers, such as the Ottoman Empire and Italy, were still instrumental in affecting the lives and even the afterlives of Tunisians like Husayn across the Mediterranean basin at the end of the nineteenth century. For instance, Ottoman representatives in Italy developed various policies and strategies for claiming authority over Tunisian citizens, pushing back against situations like those in Florence, where the French consul claimed that he was the only diplomat entitled to manage Husayn's estate.

According to article 6 of the Treaty of Bardo—the agreement that established France's protectorate over Tunisia—diplomatic and consular agents of France were "in charge of protecting Tunisian interests and nationals of the Regency" abroad.[7] In Husayn's case, French authorities needed to prove that the deceased dignitary was a Tunisian citizen under French protection. That way, they could gain access to Husayn's personal writings that threw light on his financial management of the Nessim Scemama estate, including documents considered compromising and defamatory by the French colonial authorities in Tunisia.[8] The governor of Tunisia was also interested in involving French diplomats in the conflicts over Husayn's legacy: according to Islamic law, as a former patron (*mawlā*) of the manumitted slave Husayn, the governor could inherit from him. Yet in Husayn's case, like in many others, Ottoman authorities refused to acknowledge French protection over Tunisian subjects, or even the idea of a Tunisian nationality separate from that of the Ottoman Empire. Despite French control over Tunisia since 1881, the Sublime Porte held that people from the former province of Tunis were still Ottoman subjects.

Besides France, the Ottoman Empire and Tunisia, the fourth power involved in these diplomatic conflicts, Italy, was less interested in Husayn's case per se than in related Tunisian affairs. Since its unification in the 1860s, Italy had competed fiercely with Great Britain and even with France to impose its diplomatic influence on Tunisia with the aim of eventually colonizing the country. Throughout the nineteenth century, the Italian state and its citizens developed considerable interests in Tunisia. From the 1860s until the 1930s, Italian citizens became the "largest foreign community" there. The first official census in Tunisia in 1906 recorded 81,156 Italians, 34,610 French citizens, and 1,600,000 Tunisian Muslims.[9] In Tunis, an Italian elite composed mainly of Tuscan Jews invested heavily in local trade and businesses. In the main Tunisian cities as well as in the countryside, Sicilians and Sardinians worked as artisans, miners, and farmers.[10]

These Italians and their mainland compatriots experienced the French invasion of Tunisia in 1881 as a humiliating "Tunisian slap" ("lo schiaffo di Tunis") to the Italian face.[11] From then on, Italian officials remained on the lookout for any opportunity to reassert their interest and influence in Tunisian affairs. Such an opportunity occured in December 1887, when Léon Elmilik, Husayn's former right-hand man, presenting himself as a creditor of the deceased, convinced the Italian justice of the peace to enter the chancellery of the French consulate to protect his right to attend the inventorying of Husayn's possessions.

The many conflicts over Husayn's legacy that unfolded in Florence allow us to revise our understanding of Ottoman and Italian intervention in Tunisian and other North African affairs. Until the 1980s, historians of North Africa and the Mediterranean had often argued that the Ottoman Empire and, to a lesser extent, Italy had given up intervening in Tunisian affairs. For a number of years, scholars assumed that Istanbul, which had exerted tutelage over the province of Tunis since 1574, failed to resist the French occupation.[12] Similarly, historians tried to understand the Italian government's inability to colonize a land so close to Rome.[13] Since then, scholars have devised new interpretations and asked new questions. Some studies have sought to reveal the scale of Ottoman interventions in Tunis and the Maghreb, despite European colonial domination.[14] Andreas Tunger-Zanetti's major work on relations between Tunisia and the Ottoman Empire, for instance, has brought out the multiple layers of interaction between Istanbul and Tunisian elites from 1860 to 1913.[15] On the Italian side, historians such as Juliette Bessis, Mary Lewis, Julia Clancy-Smith, and Gabriele Montalbano have explored how Italian officials exerted a sort of soft colonial power in Tunisia—and along the southern shores of the Mediterranean more broadly—by relying on Italian investors, migrants, and workers, an influence they retained until the end of the Second World War and the defeat of the fascist regime.[16] Following these recent works, the first two sections of this chapter highlight two aspects of Ottoman involvement in Italy and the Mediterranean; the third section explores the lasting impact of Italy on North African individuals and how the Italian state used litigation between them in Tuscany to contest French influence over Tunisia.

The first section contextualizes Ottoman Consul Musurus Ghikis's actions by placing them against the wider backdrop of Ottoman military and diplomatic interventions launched after 1881 in favor of Tunisian subjects. Such a proactive policy contradicts the idea of an Ottoman withdrawal from Tunisian affairs. On the contrary, it evinces a renewed Ottoman interest in North Africa. In particular, the heated debate of summer 1887 between the Ottoman consul and his French counterpart over Husayn's nationality, along with the latter's own conception of his Tunisian identity, clearly points to the fact that Ottoman and Tunisian subjects did not merely accept colonial categories of nationality. The Ottoman and French consuls never did succeed in ascertaining Husayn's nationality, but through these types of diplomatic and legal conflicts, Musurus Ghikis and other Ottoman dignitaries (including Husayn himself) developed their own conception of subjecthood and nationality. For such men, Tunisian and Ottoman citizens belonged not only to the Ottoman Empire but also to the broader community of Muslim believers (the *umma*, in Arabic).

The second section throws light on a less formal Ottoman intervention that historians have rarely explored: the administration of Ottoman subjects' deaths and funerals across the Mediterranean. In this section, I follow closely the actions of the Ottoman consuls in Tuscany in November 1887 as they exhumed Husayn's remains from the Muslim cemetery in Livorno to ship them to Istanbul for reburial in the garden of Sultan Mahmud II's prestigious mausoleum. This specific case is a vivid example of Ottoman efforts to intervene in the lives and afterlives of Tunisian dignitaries in order to claim that these men were still an integral part of this Muslim empire and its achievements.

The third and final section focuses on the last diplomatic episodes related to the conflicts around Husayn's legacy. In December 1887, three days before Christmas, an incident broke out when Elmilik obtained the cooperation of an Italian justice of the peace and *carabinieri* (armed forces) to seize documents from the French chancellery in Florence. This Italian meddling clearly went against the basic diplomatic principle of consular extraterritoriality and the inviolability of diplomatic archives. It brought about a diplomatic crisis that nineteenth-century legal scholars have labeled the "Florence incident."[17] This diplomatic event emphasizes the extent to which Italy was still a place where Tunisian and Algerian Jews and Muslims deliberately chose to exert agency. It reveals that the Italian state intervened not only in Tunisian affairs in Tunisia, but also in North African affairs involving North African and French individuals in Italy itself.

Increasing Ottoman Interest in Colonial Tunisia

Consul Musurus Ghikis's quick intervention in Florence and his fierce competition with the French consul over Husayn's legacy and nationality, from the end of June to September 1887, did not come out of the blue. The promptness with which Musurus Ghikis placed the seals on Husayn's apartment and body, on June 27, 1887, is just one of the many pieces of evidence for the Ottoman administration's increasing interest in Tunisian affairs and its ability to follow specific cases unfolding far from Istanbul.

With the French gradually establishing themselves in the former province of Tunis, Istanbul was constantly considering how to respond in order to maintain its former preeminence in Tunis and its sway over Tunisians. For instance, on the death of Tunisian governor Muhammad al-Sadiq Basha Bey in 1882, while the French were still taking control of Tunisia, the sultan and his ministers decided to follow custom and issue a *firman*, an official order of investiture for Muhammad al-Sadiq's successor, 'Ali Bey III, even if France would reject it.[18] That same year, Ottoman sultan Abdülhamid II sent representatives (*shuyūkh*) to the "tens of thousands" of men, women, and children who had left the center and south of the province of Tunis, seeking refuge in the neighboring province of Tripoli, which was still under

Istanbul's control. The *shuyūkh* (singular *shaykh*) were tasked with maintaining and galvanizing these tribesmen's feelings of defiance and resistance against the French occupiers.[19]

The placing of seals on Husayn's apartments and body was handled by individuals who moved in Tunisian Ottoman circles close to the sultan's entourage. Musurus Ghikis's efficient intervention in Florence stemmed in part from the close relationship Husayn had built up with Ottoman representatives in Italy and from the influence Khayr al-Din—Husayn's former mentor—had gained in Istanbul, where he was prime minister to the sultan from 1878 to 1879. In Florence, Ottoman consul Musurus Ghikis had been a frequent guest at Husayn's residence. Indeed, the French consul suspected Musurus Ghikis of having taken advantage of this proximity to suggest that Husayn "write his will in favor of the sultan."[20] From the late 1870s, Husayn had also been in close contact with Gallian, another agent representing Ottoman interests in Rome.[21] There, Husayn had approached Ambassador Tarkhan Bey in 1880, before the French military occupation, and came to an agreement on secret conditions pertaining to Tunisian consulates in Italy.[22]

In Istanbul, Khayr al-Din had become the sultan's trusted man, dealing with Tunisian issues on a daily basis. For instance, in 1887, he invited French representatives to confirm a governor appointed by Istanbul, thus recognizing Tunis as an integral part of the Ottoman Empire.[23] In turn, Khayr al-Din had counted on the support of Muhammad Zafir al-Madani, the head of the Madaniyya, a *sufi* or mystic brotherhood whose disciples hailed from the Hejaz and from the swathe of land running from Egypt to Tunisia.[24] Over the course of the 1870s and 1880s, Khayr al-Din's and al-Madani's interventions supplemented other actions taken by officials representing various Tunisian interests in Istanbul. For instance, throughout the nineteenth century, the Tunisian governors retained a *kapı kethüdası*, an administrator who introduced Tunisian emissaries to Ottoman ministers and ensured that official correspondence between Tunis and Istanbul ran smoothly.[25] The Tunisian administration also relied on other types of representatives, *ūkalā*, traders in charge of financial affairs who often hailed from the Tunisian island of Djerba.[26] By adding Khayr al-Din and al-Madani to the *kapı kethüdası* and the *ūkalā'*, Istanbul clearly signaled its sustained interest in the fate of the former province of Tunis and its subjects. To follow Tunisian affairs, Ottoman authorities involved diplomats such as Consul Musurus Ghikis as well as state and religious actors, combining formal and interpersonal relationships. As Tunger-Zanetti has argued, confronted "with the powerful means European states used in their efforts to control the regions they later colonized, these multiple relations were a major advantage for societies forced onto the defensive."[27]

Imperial Debates About Husayn's Nationality

On the ground in Florence, Consul Musurus Ghikis's competition with French consul de Laigue over Husayn's legacy was indeed part of a strategic move by the Ottoman Empire. By placing seals on Husayn's apartment in the hills of Florence, the Ottoman consul likely wished to bolster potential claims by Khayr al-Din and, more probably, the empire itself to Husayn's estate and private papers. Musurus Ghikis clearly intended to place Husayn's affairs under the protection of the Ottoman *bayt al-māl*, the public treasury tasked with collecting revenue from confiscated domains left unclaimed or without heir (*matrūkāt*, *mukhalafāt*).[28] But his actions were not only motivated by material interests. Neither was he simply disputing French claims to "protect" Tunisian nationals abroad. Musurus Ghikis was perpetuating a long-lasting Ottoman policy that consisted of asserting the "Ottomanness" of Tunisian citizens, including Husayn, by negating any idea of Tunisian autonomy, especially when this "autonomy" was under French control.

Beginning in the 1830s, the Ottoman consul's predecessors in Italy had fought all of the Tunisian state's attempts to be represented by Tunisian consuls separate from Ottoman representatives,[29] even though these Tunisian consulates were present on the northern shores of the Mediterranean up until French colonization.[30] For example, one of Musurus Ghikis's relatives, a consular deputy in Turin in the 1850s, had struggled to prevent the appointment of a Tunisian consul to Genoa.[31] The Ottoman Empire's emphatic negation of Tunisian autonomy was also on display in the summer of 1887, when Consul Musurus Ghikis refused to allow attorney David Santillana to act as "the official agent of the Tunisian government": an "intervention by the representative of the Tunisian administration seemed to him such as . . . to establish . . . in favor of the *bey* the status of autonomous public power which Turkey is not disposed to recognize."[32] The idea of greater Tunisian autonomy, or even of Tunisia's virtual independence, was a pernicious one for Istanbul, as it allowed

French authorities to avoid negotiating with Istanbul on how they should set up their protectorate over Tunisia.[33]

In the heated debates with the French consul over Husayn's nationality, the Ottoman consul and his superiors in Istanbul made a threefold argument that Husayn could only be treated as an Ottoman citizen. First, Husayn hailed from Circassia and, like many Circassians, became an Ottoman subject upon being brought to the empire.[34] Second, Husayn never ceased to be loyal to the Ottoman sultans: he was granted Ottoman passports (figure 4.1) as well as prestigious titles, and he had vigorously opposed the French military occupation of Tunisia since 1881.[35] Third and most interestingly, the consul kept arguing that "the acquired and recognized status" of Husayn as a Tunisian meant that the deceased was "an Ottoman subject," which was once again a way of asserting that the province of Tunis was merely a component of the Ottoman Empire.[36]

On their side, the French authorities and Santillana, acting as the Tunisian governor's representative, claimed that Husayn belonged to a Tunisian state "under French protection."[37] The French consul relied on social reputation: whether in Florence or in Livorno, "in the eyes of all, the General was viewed as Tunisian."[38] Santillana contended that in accordance with article 92 of the former Tunisian Organic Code—the constitution promulgated in 1861 and suspended in 1864—subjects of the Tunisian governor could not renounce their subjecthood, a position that Husayn himself had argued in the Scemama affair.[39] For the French consul, determining Husayn's nationality at the moment of his death was not crucial for establishing "which law to apply to the estate," because whether he was a Tunisian or an Ottoman subject, "Muslim law" applied.[40] However, ascertaining Husayn's nationality was necessary to identify which government should administer his estate in the event of there being no heir: the Ottoman state, or the Tunisian state backed by the French authorities.[41] For the French consul, at stake were not only sovereignty and "money and material loss," but also the papers that Husayn had held in his apartments.[42]

The discussions about Husayn resemble many other debates about similar cases involving Algerians and Tunisians during the colonial era.[43] In each case, diplomats and administrators had to determine if a colonized subject from the former Ottoman province of Algiers or Tunis was under French protection or Ottoman authority. Building on these various disputes about Maghrebis' nationalities across the Ottoman Empire, historians such as Will

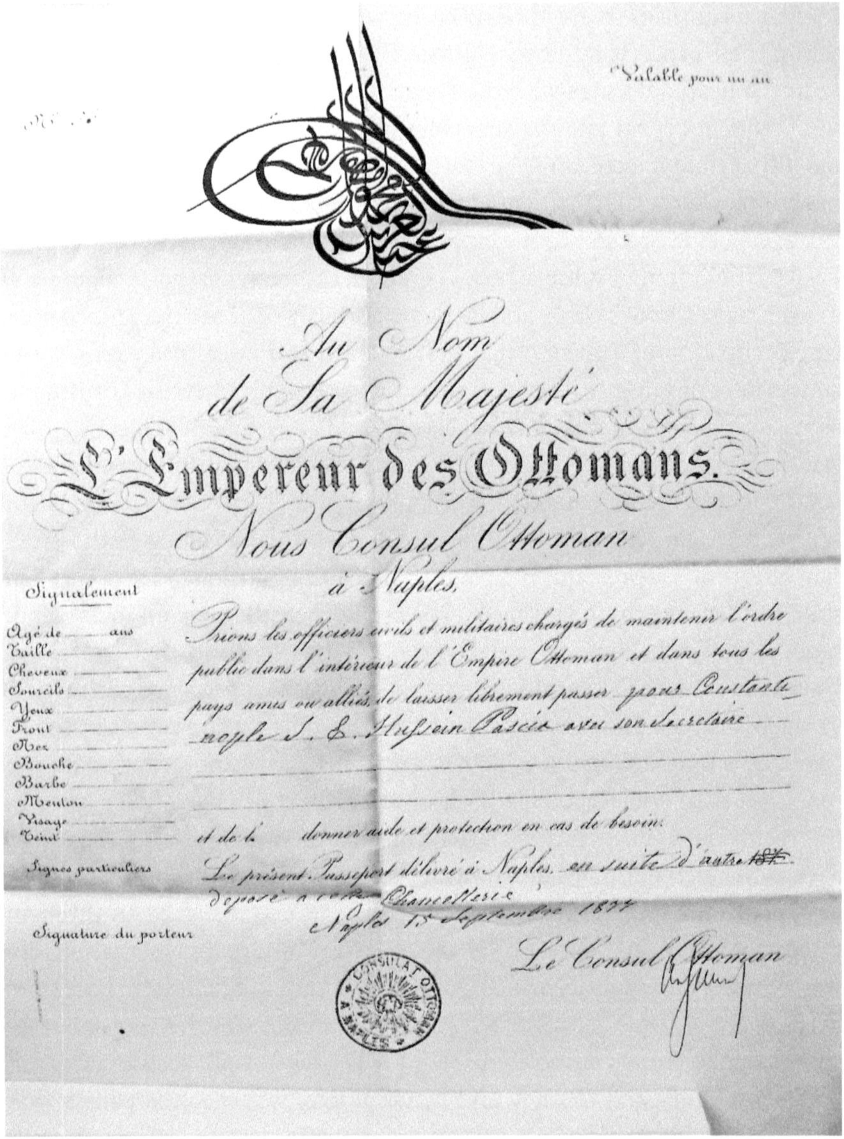
Valable pour un an

N°

Au Nom
de Sa Majesté
L'Empereur des Ottomans.
Nous Consul Ottoman
à Naples,

Signalement

Agé de ___ ans
Taille
Cheveux
Sourcils
Yeux
Front
Nez
Bouche
Barbe
Menton
Visage
Teint

Signes particuliers

Signature du porteur

Prions les officiers civils et militaires chargés de maintenir l'ordre public dans l'intérieur de l'Empire Ottoman et dans tous les pays amis ou alliés, de laisser librement passer pour Constantinople S. E. Hussein Pascia avec son Secrétaire

et de lui donner aide et protection en cas de besoin.

Le présent Passeport délivré à Naples, en suite d'autre [illegible] déposé à cette Chancellerie

Naples 15 Septembre 1877

Le Consul Ottoman

Consulat Ottoman à Naples

FIGURE 4.1 Ottoman passport issued for Husayn and his secretary in Naples, 1877. Photograph by author.

Hanley have argued that cities in North Africa such as Alexandria, in Egypt, became places "where legal nationalities of many regions and types were on offer" and developed as a matter of necessity an economy in which papers, whether "forged or genuine,"[44] were traded as commodities. This unsettled situation was in fact both an opportunity and a legal constraint for North African individuals such as Husayn. The Ottoman administration held that people from the former province of Tunis were its own subjects, refusing to acknowledge Tunisian sovereignty; in riposte to this, French authorities rarely recognized Ottoman visas and documents.

More importantly, the diplomatic discussions and negotiations about these debatable citizenships forced the Ottoman authorities and subjects to define their own understanding of Ottoman nationality. In these debates, colonized individuals and Ottoman officials were not replicating a colonial, European understanding of nationality, nor did Tunisians merely wait for the implementation of a Tunisian nationality law during the colonial period:[45] Tunisians, and Ottoman officials more broadly, shaped their own laws and ideas about nationality beginning in the 1860s, according to earlier debates about the balance of power between provinces and the imperial center.[46] As we saw, as a representative of the Ottoman Empire, Musurus Ghikis did not acknowledge the existence of a Tunisian state or Tunisian nationality, and this had an effect on his understanding of the Ottoman identity. According to the Ottoman consul and his own diplomatic hierarchy, Tunisians were members of a larger Ottoman community under the sultans' tutelage. The Ottoman identity could not be provincial or national—it had to be an imperial Islamic citizenship.

Like Consul Musurus Ghikis and other dignitaries, Husayn shared this broad understanding of Tunisian identity. Husayn gradually developed his thoughts about Tunisian nationality over the course of his involvement in the Scemama affair, starting in 1873. His most complete articulation of his understanding of "belonging" to Tunisia and the Muslim world appeared in a pamphlet he wrote five years later in Arabic and French.[47] At the heart of his thinking lay the idea of belonging to Islam and, more specifically, of being subject to a Muslim authority. For Husayn, this fundamental belonging was incumbent on Muslims and on individuals of other religions living under Ottoman authority:

> Everybody is aware that Islamic political society is based on religion (*al-siyāsa al-islāmiyya murtabiṭa bi al-dīn*), and that the rights and duties of all subjects

> (*ra'iyya*) of a Muslim state, irrespective of their race or denomination, fall under this regulation, and that they are in this dependency by virtue of their religious ties, if they are Muslim, and by virtue of the ties resulting from the pact of subjection if they are non-Muslim. . . . And both of these ties are indissoluble, for religion (*al-dīn*) forbids non-Muslim subjects from violating the pact of subjection, just as it forbids Muslims from apostasy (*al-ridda*).[48]

Husayn's thought might appear circumstantial, primarily as a way of insisting that Scemama was of Tunisian nationality. In this definition, Husayn essentialized a distinction between European countries, which gave "preference to nationalities or races" (*al-ijtimāʿ al-jinsī wa al ʿaṣabiyya*), and Muslims, who, "basing [their thought] on the fact that humanity is one," favored "religious and theocratic idea[s]."[49] Husayn's evocation of Muslim "belonging" did not refer to national borders, nor was it bounded. Like other Muslim thinkers, Husayn sought to assimilate local and provincial "belongings" together with those operating on the larger scale of the Muslim world.[50]

In the long run, Husayn's conceptions of Tunisian nationality as part of an Islamic community—as well as Consul Musurus Ghikis's advocacy for a broader Ottoman nationality—largely went unnoticed, and most of the understandings that Husayn discussed did not become law. During the colonial period in North Africa, other definitions of nationality, more centered on nation-states and their well-defined borders, overshadowed other political opinions that were based on expansive and often vague imperial and religious communities. Furthermore, such broad conceptions of "belonging" and nationality appeared to be difficult to understand and manipulate on the ground. In Husayn's case, Consul Musurus Ghikis and Consul de Laigue never managed to determine once and for all which nationality Husayn was. As early as the beginning of July 1887, the Ottoman consul in Florence confessed that he did not want to become involved "in a discussion that [his] superiors could conduct with greater authority."[51] The arguments of the French consuls and Santillana that presented Husayn as a Tunisian citizen soon started to collapse under the accumulation of contradictory evidence.[52] Nor were French and Ottoman authorities able to call on any of Husayn's relatives, given that he had been bought in Tunis as a child.[53]

The French and Ottoman authorities had no real interest in getting bogged down in legal disputes over Husayn's nationality. In particular, the main French administration in Tunis, the *résidence générale*, wanted to avoid the precedent of the Scemama affair, which had lasted for more than seven years because no one could clearly determine if Scemama was Tunisian or Italian when he passed away in 1873. After a summer of discussion, on September 4, 1887, Khayr al-Din intervened with the sultan to have the *bey* of Tunis recognized as Husayn's heir. The Ottoman Embassy in Rome henceforth had to "dispense with intervening" in this estate.[54] The seals were removed from Husayn's apartments on October 21, 1887, and the French consul was henceforth able to draw up the inventory of the estate. The nationality question reemerged one final time, in November 1887, when Husayn's body was moved from Livorno to Istanbul.[55] This was also the final occasion for the Ottomans to intervene in the Husayn affair, through their handling of bodies and mortal remains across the Mediterranean.

Ottoman Administration of Bodies and the Deceased

Husayn had been buried in the Muslim cemetery of Livorno for little more than four months—from late June 1887 until the night of November 11—when his remains were suddenly exhumed.[56] By order of the Ottoman sultan Abdülhamid II, his body was to be sent from Tuscany to Constantinople, where Husayn would finally rest in peace in Sultan Mahmud II's prestigious mausoleum.[57] This mausoleum, built in 1838, was renowned for housing the tomb of its builder—who implemented a new wave of Ottoman military and administrative reforms beginning in 1826—and the gravestones of his successors, Sultans Abdülaziz and Abdülhamid.[58]

The exhumation of Husayn's remains and his second burial in such a prestigious building are striking and surprising. Husayn was clearly a major dignitary in the Ottoman province of Tunis. However, at that time, he was one of the few, if not the first, provincial Tunisian dignitary to be buried in an imperial mausoleum in Istanbul. Usually, local dignitaries such as Husayn would be buried in Tunis, either in the major cemetery of the Jallaz or in the *turbat al-bey*, a local mausoleum of the Tunisian governors.[59] What triggered the Ottoman sultan to act at this precise moment and exhume Husayn's remains? In fact, unlike most of his peers, Husayn could not be buried in Tunis. The French clearly rejected Husayn, who had been a staunch opponent of the French and their efforts to dominate Tunisia since the 1870s. Furthermore, while Muslim norms impose rapid burial, the

Tunisian governor hesitated a long while—too long—over whether to have the body of his former servant repatriated.[60] By late June, Husayn's body would not have been able to withstand the extended journey to other Mediterranean shores. General Husayn was buried on June 30, three days after he passed away, without the specific honors due to his military rank.[61] That first funeral was announced in the Italian press on July 7, ten days after he had died.[62]

The solution of temporarily burying him in Livorno probably enabled Angiolina Bertucci, Husayn's lover with whom he had lived in the city, to mourn him. Husayn lay in the Muslim cemetery for more than four months, a period during which, according to Muslim norms, widows may not remarry.[63] In late October, Ali Bey, the governor of Tunis, finally ordered Santillana "to send" the body to Tunis,[64] but financial reasons eventually dissuaded him from this course of action.[65] It is only at this stage that the Sublime Porte decided to claim Husayn's remains. In a memorandum, the Office of the Grand Vezirate argued that the "deceased had made it openly known while alive that he wished to be buried in Istanbul."[66] The Ottomans could thus reassert their authority over a Tunisian notable through "necropolitics"—a term I borrow from Achille Mbembé, not in his original meaning of the "power and the capacity to dictate who may live and who must die," but in reference to the administration of Ottoman subjects' deaths and funerals across the Mediterranean.[67]

Historian Mary Lewis has shown to what extent Tunisian nationalists used social and political conflicts around burials and cemeteries to define Tunisia as a Muslim nation in the early 1930s. At the time, nationalists prevented Tunisian citizens who had opted for French nationalization from being buried in Muslim cemeteries.[68] But well before the interwar period, in the mid-1880s, issues surrounding cemeteries had triggered other political demonstrations in Tunis. In 1885, for instance, a member of Husayn's entourage, Muslim scholar Muhammad al-Sanusi, drew up a petition in reaction to new municipal regulations on burying the dead.[69] Two years later, in 1887, Jewish families protested as French authorities decided to close the old Jewish cemetery in northern Tunis and convert the land to public use.[70] These families could only use this place to temporarily house the bodies of their relatives while waiting for them to be carted to the new Jewish cemetery at Borgel.[71] These two protests did not result in major political mobilizations in Tunisia. However, they clearly demonstrate that cemeteries were becoming sensitive, contentious places for identity politics in the colonized state.

Cemeteries and burials became sites of concern for Tunisian and Ottoman subjects abroad for a second reason. By the mid-nineteenth century, Ottoman officials like Husayn, who were sent to Europe on diplomatic missions and decided to settle in European countries, started to be concerned about their own potential burials in Christian lands.[72] Writings by Husayn and his intimates refer to these burials and how to respect Muslim rituals in Europe.[73] For instance, in his letters to Khayr al-Din, Husayn describes the funeral of an eighty-three-year-old Persian man in London, who was buried in the Muslim plot next to the Jewish plot, "on a parcel of land bought by Emir 'Abd al-Qadir."[74] And in his travelogue, Husayn's friend al-Sanusi also refers to the modest burial of 'Ali b. Muhammad b. Jabir, who had come to Tuscany in Husayn's service and died there in June 1882.[75]

Ottomans were aware of these concerns. They gained significant expertise in dealing with the death of their subjects outside the empire and across the Mediterranean. In Husayn's case, the Ottoman consul in Florence was—as we saw—the first to place seals on Husayn's body, even before the French consul.[76] More importantly, the Muslim cemetery of Livorno, founded outside the city in 1762 by traders from Tripoli and officials from Algiers, was completely under Ottoman control by the 1880s.[77] To inter Husayn there, the Ottoman consul in Livorno had to employ gravediggers, stonemasons, and workers to build Husayn's tomb, as well as guardians "to watch over" his "corpse."[78] The Ottoman consul also had to adapt Muslim rituals to the new hygienic norms that were in force in this part of Europe[79]— bodies were to be washed outside of town, for instance.[80] The Livorno cemetery clearly shows that even though Ottoman authorities had to renounce their claims to territory in Algiers, Tunis, and eventually Tripoli, they took over certain responsibilities concerning North African representatives in Europe.

But Ottomans were not only appropriating the Tunisian governor's responsibilities over his deceased subjects. The central Ottoman government decided to transfer Husayn's remains in order to claim that Tunisian dignitaries were still an integral part of the Ottoman Empire and its achievements. Husayn's remains were sent to Istanbul at the instigation of Khayr al-Din by mid-November 1887, when the disputes over Husayn's nationality were more or less settled.[81] On this occasion, the French authorities again expressed their fear that the sultan's men might conclude from this transfer that Husayn was of Ottoman nationality, thereby reigniting claims to Husayn's estate.[82] Their worries dissipated once the work of inventorying papers in Husayn's estate began in the buildings of the French consulate. The possibility of letting the body pass through France was even considered, as the Austro-Hungarian authorities had refused to allow it transit through their territory. In the end, Husayn's remains were transported to Marseille, where they were stored in a lead and zinc box that was labeled "antiques."[83] From there, the box was sent by sea on the steamer *Avlona*, belonging to the English company Thomson. It arrived in Istanbul by late November 1887 under the escort of ʿAmr Bu Hajib, the son of Husayn's secretary, Shaykh Salim Bu Hajib.[84]

The fact that Husayn's remains were buried a second time in the garden of Sultan Mahmud II's prestigious mausoleum was meaningful. Mahmud II had promoted a new type of army trained and equipped along European lines (*al-niẓām al-jadīd*), together with other administrative reforms launched in 1826. As seen in the first chapter, not only did Husayn benefit from these reforms, but he was also deeply involved in their implementation. As a matter of fact, Husayn's funerary inscription in Ottoman Turkish in this mausoleum links him to the Ottoman state reforms, specifically mentioning his appointment to the Ottoman council of state (figure 4.2).[85]

But Ottoman "necro-politics" went beyond the relocation of a Tunisian citizen to a dynastic building exalting the main actors of the reforms (*tanẓīmāt*) across the empire. Public funerals in the major Ottoman cities became significant moments of political expression. Individuals who took part in the funerals of Ottoman Maghrebis expressed a sense of Muslim solidarity, affirming the ties binding North African subjects to the populations and territories of Islam.[86] Although we have found no descriptions of Husayn's second funeral in Istanbul, we know that during the same period, in 1889 in Cairo, a "sad and weeping crowd followed" the funeral of Husayn's compatriot Muhammad Bayram V, who left Tunis after the French colonized his country.[87] The following year, "the funeral of Khayr al-Din Pacha," friend of Husayn ibn ʿAbdallah, former Tunisian prime minister, and then vizier of Sultan Abdülhamid II, was celebrated with some ceremony in Istanbul before being placed in a special vault in the Eyüp Mosque where other Ottoman dignitaries were buried.[88]

FIGURE 4.2 Husayn's tomb at the mausoleum of Mahmud II in Istanbul. Photographs by Nadirah Mansour.

Ottoman subjects also used not only the streets and public spaces but also newspapers to praise those North Africans who maintained their faithfulness to the Ottoman sultan. The press lionized these deceased servants and described them as champions of the Muslim world. When Husayn died, Bayram V penned an obituary in *al-I'lām*, the newspaper he had founded in Egypt, titling it "National Tragedy" (*Fāji'a wataniyya*) and portraying Husayn as a hero of Islam.[89] Bayram V glorified Husayn's passion and zeal for defending the Muslim world and helping martyrs of Islam with "his body and his goods."[90]

Both this article and an obituary in another newspaper, *al-Majalla*, praised Husayn's involvement in the Crimean War (1853–56), in which the Ottoman Empire—allied with France and Great Britain—fought Russia, mainly on the Crimean Peninsula. They mentioned that the deceased dignitary had cried on reading newspaper reports from the front and asked to enlist. Husayn is said to have helped the wounded "as if he were a member of the Red Crescent."[91] Tunisian governor Ahmad Bey sent more than ten thousand troops to support the Ottoman military effort from 1854 to 1855, and General Rashid, a mamluk officer, took command of the Tunisian army. We have no indication that Husayn took part in this expedition at all.[92] The *Majalla* article also reported on Husayn's role in educating Tunisians and his wish to see them become engineers and doctors. It emphasized his support for the attorneys defending Urabi Pacha, the leader of the uprising against French and British forces in Egypt in 1881–82.[93] Bayram's article made no reference to Husayn's settling in Italy; it simply stated that Husayn was traveling between Italy, France, and England, and that he never missed a prayer wherever he went.[94]

The authorities' decision to bury Husayn in an Ottoman mausoleum, the mobilization of crowds for funerals, and the publication of obituaries in the press were all ways for Ottomans to assert their political power.[95] Exploring all these layers of Ottoman policy—ranging from the placing of seals to the administration of deaths and burials—allows us to underline the major fact that Ottoman and then national political cultures continued to use burials as mass political events in the Maghreb throughout the colonial period.

Once Husayn's body had found its final resting place in the mausoleum of Mahmud II, Ottoman authorities no longer intervened in any notable manner in the affair over his estate. In late 1887, it was Italy's turn to play a brief role in the conflicts over Husayn's legacy.[96]

Italy and the Diplomatic Incident in Florence

Three days before Christmas 1887, at half-past noon, a justice of the peace (*praetor*) from the first judicial division of Florence knocked at the main door of the French consulate, armed with a ruling from a local court and escorted by a clerk, a bailiff, and two lawyers. In the absence of the French consul, the judge decided to force his way into the chancellery. There, challenged by Chancellor Langlade, the judge brought in armed *carabinieri* and "ordered a locksmith to open the door" to a room where the Tunisian government's interpreter-chancellor and archivist, Chekri Ghanem, was reading and inventorying Husayn's papers. Despite Ghanem's vehement protests, the judge seized documents "that he presumed to belong" to Husayn's estate. After forcing open a large archival cabinet, the Italian officer placed his seals on other papers and ordered forced entry to another room where he seized other writings and left "an entire shelf of state papers" in total disarray.[97]

Once the justice of the peace left, officers at the consulate started to draw up a report,[98] but the employees were still in a state of shock. The Italian justice's intervention at the French consulate triggered a noteworthy diplomatic incident. After a meeting at the British consulate on December 25, 1887, the British, Greek, Russian, and American consular authorities in Florence unanimously denounced the unjustified violation of basic diplomatic principles of the extraterritoriality of the consulate and the inviolability of its archives.[99] The French and Italian newspapers dramatized what was very soon referred to as "the Florence incident."[100]

Mainstream newspapers never missed an opportunity to inform their readers of the tensions and animosities between the French and Italian nations. The two neighboring countries were not only competing over Tunisia at the time; they were also in the midst of yet another customs war after "the adoption of protectionist tariffs" in 1887, and the press was not helping.[101] Nearly a month after the "incident," an Italian newspaper, the *Unione*, reported on a similar event that took place on January 25, 1888 in Damas, in which an Ottoman police officer accompanied by fifteen armed policemen broke into the French consulate to drive out an Algerian Muslim.[102] Two international law reviews even decided to publish long articles analyzing the legal consequences of the "Florence consulate incident."[103]

All the fuss caused by the diplomatic incident led the French authorities to forget the person responsible for this consular disaster in the first place, the Algerian Jew and naturalized Frenchman, Léon Elmilik. Husayn's former right-hand man, having presented himself as a creditor of the deceased and as a representative of the Tunisian governor, provoked the Italian intervention when the French consul excluded him from the inventory of Husayn's estate taking place within the French consulate. Elmilik complained to a Florentine justice of peace, who ruled against the French consul and decided that, as one of his creditors, Elmilik was entitled to be involved in inventorying Husayn's estate.[104] One month later, in late January 1888, the Florence consulate incident was resolved, and Elmilik was allowed to attend the inventory in the archives of the French chancellery.

Historians of Italy and North Africa have rightly assessed the political and diplomatic effects of the mass presence of Italians in Tunisia and the Maghreb.[105] Elmilik's crucial role in triggering this diplomatic crisis is fascinating, however, for what it reveals about the ways men and women from the Maghreb acted in Italy once they had settled there and how Italian officials could use these North Africans for other political agendas. Elmilik's involvement helps us think of Italy and other European countries not just as sovereign powers engaged in international rivalries, or even as mother countries for their migrant communities during the colonial period, but also as arenas for action by North African citizens.

As we shall see in the next chapter, other actors involved in the Husayn affair also used Tuscany as an avenue for their disputes over Husayn's estate. These people, who originated from or had interests in the Maghreb, were less numerous than the Italian communities residing in North Africa. Still, we need to understand how and to what extent men and women from the southern shores of the Mediterranean—or with sizable interests there—could draw on the Italian diplomatic and judicial systems or, more generally, on institutions in countries other than colonial France or the Ottoman Empire. In Elmilik's case, as in others, Italy and other European countries were home to institutions and extraterritorial places (such as the Muslim cemetery in Livorno and the archives in the chancellery of the French consulate in Florence) that contributed to fashioning the modern Maghreb.

We should not consider Italy or other places outside of colonial North Africa only as "sites of exception," as "third places" that offered particular legal opportunities to North African subjects.[106] These colonized subjects could be at the Italian state's mercy, just as Italian officials could make use of

North African legal conflicts or, on the contrary, ignore their claims entirely, depending on the state of relations between Italy and France. Along the same lines, the Ottoman Empire was not only providing military and diplomatic assistance and funerals to its former Algerian and Tunisian subjects across the Mediterranean. As we have seen in this chapter, Ottoman officials were consciously and vigorously contesting both Husayn's provincial, Tunisian identity and the existence of Tunisian diplomatic representation in Italy. In a period of European colonial expansion, the Ottoman authorities' goal was obviously not to emancipate or decolonize their North African subjects but increasingly to "re-Ottomanize" them, to reassert the sultans' authority over North African populations whose nationalities and legal statuses were more uncertain than ever.

In this sense, reappraising the history of Tunisians and, to a certain extent, Algerians in a Mediterranean context—not just in the context of French colonial domination but in relation to other regional powers such as Italy and the Ottoman Empire as well—not only results in recovering other sites of actions and opportunity for North African men and women; it also means taking into account the many challenges these populations faced while finding their own paths amid the interests of the various hegemonic powers in the Mediterranean. Crucial to our story is the idea that throughout the conflicts over Husayn's legacy, states and empires were not the main actors. The Ottoman authorities, as we saw, intervened only in the first stage, from late June to late November 1887, and the Italian justice only assumed a crucial role around Christmas 1887 with the "Florence incident." As we shall see in the next chapter, individuals, such as Elmilik, the Bu Hajib family, the governor of Tunisia, and the mothers of the two daughters educated by Husayn, played influential roles for a longer period of time—until the beginning of the 1890s, and even up to the 1920s in the case of the Bu Hajib family.

FIVE

Sovereigns, Mothers, and Creditors

The Agency of Husayn's Potential Heirs

ON MANY occasions, all the parties interested in Husayn's estate, or their representatives, found themselves squeezed into the archives of the French chancellery: Léon Elmilik and other North African Jews who had been Husayn's creditors in Tuscany; lawyers representing the interests of the two girls raised by Husayn in Livorno; David Santillana as a spokesperson of the governor of Tunis; translator Chekri Ghanem; and the French chancellor. These men would gather on a regular basis to attend the inventorying of the deceased's papers from fall 1887 to spring 1888. From time to time during these meetings, Elmilik would loudly express his dissatisfaction, shouting angrily and even breaking into tears in the French consulate.[1] Santillana—raised in Tunis in the entourage of the British consul and educated in Victorian England— reacted to Elmilik's lack of decorum and sophistication with profound scorn; and Ghanem, alongside this pair, was a less than confident translator of Arabic to French.

Each of these gatherings was extremely tense. The claimants and their lawyers hoped (or, conversely, feared) that the translator and the chancellor would find new written evidence in Husayn's papers that would support or ruin clients' claims to Husayn's legacy. Each meeting and each reading or translation of Husayn's papers from Arabic into French brought its share of surprising documents and new developments. One of the most contested documents was Husayn's secret will, written in Italian, which was made public outside of the French consulate in Florence on July 2, a few days after Husayn passed away.[2] Muslim law held that Husayn, as an emancipated slave without recognized descendants, was to pass his assets to his master and his master's descendants—that is to say, to the governor of Tunis, 'Ali Bey, and his lineage.[3] But in his will, Husayn followed other Islamic law principles that allowed him to bequeath one-third of his assets to Emma/Amina (then age eleven) and Maria/Myriam (then age four), the two girls who were not his official heirs but whom he has raised in his household, provided that on reaching adulthood they married Muslim men.[4] Husayn's will did not entirely preclude 'Ali Bey's rights to his assets, but it sought to diminish the Tunisian governor's claims. Consequently, during each meeting at the consulate, Santillana (as a representative of 'Ali Bey) aimed to find documents that would weaken the provisions of Husayn's secret will.

Another document that Husayn had signed, a deed passing the management of his properties in La Goulette after his death to the sons of his secretary Salim Bu Hajib, prompted other discussions and litigation. In the mid-1860s, Husayn had already transformed these properties into *ḥubūs* (or *waqf*), a charity trust that directed the income they generated first to private individuals (*ahlī*, in Arabic) and then, ultimately, to pious works of public utility (*khayrī*, according to the Islamic legal vocabulary). One year before passing away, Husayn designated invalid soldiers from the Ottoman army as the long-term beneficiaries of his *waqf*.[5] But his desire to favor the Bu Hajib family in the short term raised questions in 'Ali Bey's entourage. Husayn sought to reward Bu Hajib's sons for their help during his exile in Italy.[6] However, by doing so, Husayn deprived his master and the latter's descendants of yet another sizable portion of their inheritance, amounting to nearly a quarter of his fortune.[7] The only thing the manumitted slave left to 'Ali Bey was the palace in the old city of Tunis that the French army had been renting since 1881. More notably, according to the Muslim law, as the legitimate heir, 'Ali Bey was obligated to pay off Husayn's debts to his creditors—including to Léon Elmilik.

These conflicts between 'Ali Bey, Bu Hajib's sons, the mothers of the two girls, and Husayn's many creditors, from the summer of 1887 through early 1890s, will be at the center of this chapter. By following the claims raised by the various interested parties in Tuscany—a land that did not depend on French colonial authorities or fall within the legal jurisdiction of the Ottoman Empire—I seek to recover each individual's specific agency according to that person's social class, gender, and/or religious identity. In this chapter, I contend that each individual's adaptability allows for an emphasis on various aspects of history that have been overshadowed in the modern and colonial history of Tunisia, such as the lasting social and political significance of Muslim households. As theorized by supporters of subaltern studies and political scientists such as Romain Bertrand, colonized subjects' practices and their "modes of understanding . . . were not exclusively, or even primarily, finalized by the deeds and actions of Europeans"; they could also "elude the control and attention of official and non-official agents of colonial power."[8] In other words, each of these historical actors reminds us that the specific places and social situations that were instrumental to exerting agency during the Ottoman era were still crucial after the advent of French colonization in Tunisia. Taking them into account allows us to shed new light on the beginning of colonization in Tunisia.

'Ali Bey, the two mothers, the Bu Hajib family, and Léon Elmilik belonged to four groups that have interested historians of modern North Africa: the Ottoman provincial elite, European women, Muslim scholars, and North African Jews. In the sections that follow, I will present the claims of these four types of actors and discuss their specific agency in light of current debates and literature devoted to each group. Starting with the case of the Tunisian governor 'Ali Bey, the first section shows that even under French control, Tunisian sovereigns were not puppet rulers shorn of power by colonial conquest and the progressive rationalization of their bureaucracy, as has often been assumed in the historiography on colonial Tunisia. 'Ali Bey's many actions to defend his rights over Husayn's legacy show that this monarch retained some ability to maneuver.[9] In order to keep some parts of Husayn's estate, including his palace in Tunis, under 'Ali Bey's control, French authorities needed to assert the Tunisian governor's rights. In particular, they could not rescind or deny his legal rights as Husayn's former master (or *mawlā*). Moreover, 'Ali Bey not only acted through his administration and other official channels but also relied on personal relations both inside and outside his palace; as a result, not all his actions were documented, posing some difficulties for historians.

In the second section, I will compare the claims and agencies of the Italian Angiolina Fortunata Bertucci and the German-French Eva Keusch, the mothers of Husayn's two heirs. I will show that Eva did not manage to successfully defend her daughter's interests even though she was a French citizen and could theoretically count on the support of the French authorities. Angiolina, on the other hand, was far more effective in defending her daughter's claims, first in Italy and even more so in Tunisia by relying on the support of the Bu Hajib family and later by becoming part of their household in Tunis. The fascinating case of Angiolina and her daughter suggests a different way of interpreting the actions of European women in North Africa—namely, that Muslim households, such as that of the Bu Hajib, which colonial authorities viewed as conservative, could in fact empower European women of less fortunate backgrounds like Angiolina. Such households were sometimes more favorable places from which to act than those available within colonial networks.

The Bu Hajibs were a third sort of protagonist, the administrators (as beneficiaries) of the income generated by Husayn's *awqāf* in La Goulette. The Bu Hajibs were Muslims endowed with cultural capital and strong ties to the circle of Muslim scholars (*'ulamā'*) based in Tunis. In the conflicts over Husayn's legacy, the Bu Hajibs had to prove their right to control this sizable portion of his estate that 'Ali Bey claimed. To make their case, 'Amr, the eldest son of the shaykh Salim Bu Hajib, penned a defense of his family's rights that portrayed 'Ali Bey as subject to the law, just like any other citizen. The Bu Hajibs' dispute with the sovereign of Tunis contributed to reformulating the relationship between the circle of *'ulamā'* and the governing sphere in Tunis at the beginning of the colonial period.

Among the creditors, the Elmilik cuts a singular figure. He allows us to explore another sort of agency—that of an Algerian Jew with French nationality who managed to distance himself from all the communities he had come from, together with the entire set of Tunisian, Italian, and French circles in which he moved while in Tuscany. Indeed, Elmilik built up a string of loyalties with French and Tunisian patrons and then defied them in order to expand his agency and defend his claims over Husayn's legacy. Elmilik's example confirms recent ideas in current scholarship that Jewish mediators like Elmilik were still part of "transregional" networks across the Mediterranean. More importantly, it demonstrates that French Jews were not all loyal to France—some of them could (and did) challenge colonial authorities.[10]

'Ali Bey's Clandestine Actions

Husayn's official and universal heir, 'Ali Bey (1817–1902; figure 5.1) was sixty-five years old in 1882, when he succeeded his older brother, Muhammad

FIGURE 5.1 'Ali Bey (Nationaal Museum van Wereldculturen, TM-60028516)

al-Sadiq Basha Bey, becoming the thirteenth governor of Tunisia from the Husaynid Dynasty (founded in 1705 by Husayn Ibn 'Ali).[11]

French authorities in Tunis constantly attempted to control the Tunisian governor's actions, even in the matter of 'Ali Bey's claims to Husayn's estate. The French got to choose Husayn's estate executor in Tunisia: Théodore Proust, a close acquaintance of Justin Massicault, the *résident général* or chief French official in Tunisia, and head of the French administration's finances in Tunisia.[12] After merely consulting the Tunisian prime minister, the protectorate's officials also appointed David Santillana as representative of the Tunisian sovereign in the Husayn affair in Italy.[13]

Santillana (1855–1931) was not a French citizen. Neither was he meant to defend French interests in Tunisia. Born to Moses Santillana, an interpreter at the British consulate in Tunis who became a British subject, David Santillana was raised in Tunis with the children of British consul Richard Wood. He was then educated in London, but left in 1873 at age eighteen for a position as an interpreter in the Tunisian Foreign Affairs Office in Tunis and then as a secretary at the International Finance Commission, which had been established in 1869 to manage Tunisian public debt. He later quarreled with French authorities over a parcel of land located in Sidi Tabet, to the north of Tunis, and was eventually forced to resign from his Tunisian positions in 1879. Santillana then decided to move to Rome, where he studied law. Despite this contentious past, in 1887, French officials chose him to represent 'Ali Bey for at least three reasons: in Florence, where he had settled, Santillana was, by and large, considered an honest lawyer; he was fluent in French, Italian, Arabic, and English; and, most important, he was intimately familiar with Tunisian affairs in Tuscany and had even worked for Husayn in 1883.[14]

Hand in hand with the French administration, Santillana did his best to curtail 'Ali Bey's claims to Husayn's estate in favor of the interests of the Tunisian state. He and the French officials first sought to distinguish between the *bey*'s private interests and the public interests of the Tunisian state in Husayn's legacy. Any contention between the *bey* and his administration that might tarnish the "prestige of the Tunisian administration" had to be avoided.[15] For Santillana and his French collaborators, 'Ali Bey was a legitimate *private* heir of Husayn. But, as the Tunisian governor and head of a state that had been bankrupt since 1867, 'Ali Bey had to collect all the wealth he could from Husayn's estate in order to contribute to the French administration's "efforts" to reduce Tunisia's debt to its creditors, per article 7 of the 1881 Bardo Treaty establishing the French protectorate over Tunisia. Santillana gave 'Ali Bey no choice. If, as Husayn's heir, he was unable to repay the deceased's debts, the Tunisian government would have to "advance them in his stead," and "if the *bey* does not agree," then he would "deserve no clemency"—"we [the French] will seize it from the civil list," the allowance given by French officials to the governor and his family for their expenses.[16] Santillana was firmly convinced that 'Ali Bey had to be coerced into cooperating. Santillana urged the French to "believe [him as] someone who knows Tunisia and the Tunisians well [because we] have a moral undertaking towards the creditors" of the Tunisian debt.[17]

At the same time, Santillana was ill at ease with the legal circumstances surrounding the estate—more precisely, the fact that Husayn, as a manumitted slave, passed on part of his estate and debts to his former master and patron, 'Ali Bey. Perhaps because he had been exposed to abolitionist ideas while educated in Britain, Santillana was deeply concerned by the continued reference to slavery in the Husayn affair. He tried to convince French officials that the protectorate should put an end to "the most deplorable traditions of the former [Tunisian] regime" so as to avoid any "further scandals."[18] Santillana might have been sincere in his claims that any kind of institution related to slavery in colonial Tunisia had to be abolished. In any case, his moral convictions weakened yet another of 'Ali Bey's rights: his total authority as a slaveholder over his remaining African slaves and Caucasian mamluks as well as over their legacies.

Why the French Needed 'Ali Bey and the Institution of Slavery

While Santillana and his French partners had ambitious plans for Tunisia and its sovereign, they could not fully disentangle 'Ali Bey's private interests from those of the state, nor could they—even if they wanted to, which they did not—get rid of certain legal institutions related to slavery. On the first point, public and private interests were deeply intertwined, although Ottoman officials in Tunisia were not ignorant of the differences between the two. For instance, the state tax and finance registers from the eighteenth and nineteenth centuries made many distinctions between state servants and the Tunisian governor's family. During the same period, modifications to the architecture of Ottoman governors' palaces led to the separation of the princely harem from those sections open to Tunisian nationals. But in the case of Husayn's estate and legacy, as seen in chapter 2, public and private property overlapped. Husayn's palace in the old city of Tunis had initially belonged to the Tunisian state, and Husayn had received the lands located in Halq al-Wadi from an earlier Tunisian governor, Muhammad al-Sadiq Basha Bey.

As to the second point, the colonial authorities could not simply abolish institutions related to slavery such as the *walā'* (the contractual relationship between masters and former slaves). They desperately needed this institution to get control over the considerable estates of manumitted slaves like Husayn, who were prominent officials before the colonial period in Ottoman Tunisia.[19] A Tunisian who had "long resided in Istanbul" and was an anonymous informer for the French ambassador to the Sublime Porte felt that Santillana's suggestion to do away with *walā'* was counterproductive:

> Requesting the *bey* of Tunis to henceforth relinquish exerting the rights he holds under Muslim canon law regarding mamluks, be they emancipated or not, would place incalculable sums in the service of our enemies [the Ottoman government in Istanbul who could claim that the mamluks were Ottoman subjects]. But by getting the properties owned by mamluks, these assets would no longer elude the French government's close surveillance.[20]

Recognizing 'Ali Bey's rights to the Husayn estate made it possible, for example, to maintain control over one of the strategic components among his assets: the palace (the *dār Ḥusayn*) standing at one of the main entrances to the medina in Tunis, which the French army had transformed into one of its main barracks in the city in 1882.[21]

More important, the French could not control 'Ali Bey's actions, his movements, or whom he met with. In the Husayn affair, the Tunisian governor acted without consulting the French authorities in order to assert his rights as legal heir in all jurisdictions—in Tunis, of course, but in Cairo and Florence as well.[22] In summer 1887, 'Ali Bey appointed his son Mustafa Bey as trustee to the estate in order to recover monies that Husayn had lent to Ibrahim al-Sanusi, a Moroccan citizen who had resettled in Egypt.[23] Beyond that, Mustafa Bey recruited Elmilik to represent his father's interests in Tuscany, despite Santillana's having been appointed to a similar role by the French and Tunisian authorities.[24] When the French authorities alerted 'Ali Bey and his son to Elmilik's record and pressured them to sever all ties with him, 'Ali and Mustafa Bey stated that they were unaware of Elmilik's ongoing problems with the Tuscan courts.[25] It took five months for 'Ali Bey to disavow Elmilik in a *beylical* decree. Elmilik then kicked up a scandal, provoking the diplomatic incident in the French consulate in Florence.[26] Despite the official disavowal, pronounced in December 1887, Mustafa Bey maintained his business relations with Elmilik.[27] Moreover, the agreements with Elmilik were reinstated in summer 1889. On another occasion, 'Ali and Mustafa Bey mobilized support in Istanbul, despite the French administration's wanting to avoid any contact between the Bey dynasty and the Ottoman government. In February 1888, Santillana suspected the French ambassador's anonymous informer based in Istanbul of actually working for the *bey* and his son.[28] Seven years later, the central police station in Tunis reported that Mustafa Bey was still in touch with Istanbul after he had sent an emissary there.[29]

Historians have shown the extent to which French authorities sought to monitor each and every aspect of the Tunisian governors' political life: their ceremonies and rituals, their civil list and incomes, even the order and practices of their succession.[30] However, all of these unsanctioned actions and hidden connections shed light on 'Ali Bey's individual agency. Thus, the Husayn affair shows that the Tunisian sovereign, even under French domination, could still act by secretly using state resources—drawing on practices prevalent throughout the nineteenth century—while also relying on family, mainly his son, and former servants such as Elmilik.

By exploring such capacities for agency at the boundary between administration and dynastic household, we can gain a far better understanding of how a North African sovereign acted under colonial tutelage. If we were assessing 'Ali Bey's ability to act based on the clauses of the treaties establishing the protectorate, or the French administration's wishes to distinguish between the public and private spheres, then such agency would appear minimal. If, on the other hand, we take into account issues that were still crucial for this sovereign—his family life, interpersonal relations, and management of real estate in Tunisia—then 'Ali Bey's interventions appear more diverse and slightly more significant. Currently, we have only a limited idea of Tunisian sovereigns' abilities to act during the colonial period, given the assumption that the once mighty *beys* were marginalized and their authority depersonalized in an inexorable march toward bureaucratic rationality.

Two Mothers, from Florence to Tunis

As universal heir to Husayn's estate, 'Ali Bey had to face various rivals, including Eva Keusch and Angiolina Fortunata Bertucci, the mothers of two girls who were to be granted one-third of Husayn's assets according to Husayn's secret will. We find scant information in the primary sources about the respective ages and educations of these two women; nevertheless, what little we have found in the French, Italian, and Tunisian archives suggests that they came from humble backgrounds.

Eva never ceased moving around from one place to the next. Born in the southwestern region of the German Empire, she followed her parents to French Algeria in 1853; later on, in the 1870s, she was hired in Florence as a "housekeeper" in Husayn's household.[31] After leaving her daughter Emma/Amina with Husayn in 1877, Eva severed all ties with Tuscany. She went to North America, where, in 1880, she married Bertrand Meuville, a New York businessman living in Montreal.[32] Because of her many travels, which brought her in contact with Arabic-, French-, or Italian-speaking civil servants, the spellings of both her first and family names were highly unstable; the last name in particular fluctuated between Heusch, Menville, and Minville.[33] Variations on her name outnumbered those of Angiolina Bertucci, who was also called Fortunata Bertucci and sometimes even Fortunata Kūnshī in Arabic, indicating a confusion with Eva Keusch's family name.

Even less is known about Angiolina. To a certain extent, like Eva, Angiolina seems to have been quite isolated. Her death certificate indicates the names of her parents, Giovanni Bertucci and Ama Manelli.[34] But no other document refers to family or even any friends who could have helped her in Livorno and Tuscany. To support her daughter's inheritance claims, in 1887, Angiolina sought the assistance of a Tunisian citizen, 'Amr Bu Hajib, eldest son of Shaykh Bu Hajib, who had been Husayn's secretary in Florence in the 1870s. In Italy, Angiolina and 'Amr became intimate, even going to a spa town together in summer 1888.[35] Husayn designated 'Amr as the executor of his will. It was he who would manage Emma/Amina and Maria/Myriam's estates and monitor their education until the two girls were old enough to decide if they would respect an important clause of Husayn's will: the stipulation that on reaching adulthood they marry Muslim men.[36]

Both women had been Husayn's employees. Perhaps they knew one another. Indeed, Angiolina raised Eva's daughter after the latter left for North America, and Husayn raised Emma/Amina "with the intention of being able to marry her off."[37] But it is that the two women developed a strong relationship prior to Husayn's death: Eva worked for Husayn in Florence until the late 1870s, and it was not until the 1880s that Angiolina took charge of Husayn's household in Livorno, where she gave birth in 1883 to Maria/Myriam. Not until the conflicts over Husayn's estate in the summer of 1888 did the two women agree on a strategy to defend their daughters' interests.

Eva initially hesitated over how to press her claims before the Italian courts and French consular authorities.[38] By the end of 1887, as the legal representative of her daughter's interests, she had asked only for alimony from 'Ali Bey and, by April 1888, for an additional 1,500 French francs as well as a "small amount of money to travel to Tunis."[39] Angiolina's lawyers asked for a minimum payment of ninety thousand French francs—with interest—from the Tunisian government,[40] and under Angiolina's influence, Eva started to put forward more ambitious demands on Emma/Amina's behalf. In the summer of 1888, the women caught the attention of Santillana and French consular authorities by threatening to prevent papers in Husayn's estate from being sent to Tunis.[41]

However, the alliance between the two women soon fell apart. Under pressure from the French and Tunisians, Eva cracked. Santillana complained about her "stupidity,"[42] although in their messages to the *résidence générale* in Tunis, both Santillana and French bureaucrats had to come up with a variety of ways to convince her to compromise.[43] Eventually, Eva accepted a contract that was "less favorable than that of Bertucci."[44] At the invitation of 'Ali Bey, Eva agreed to send papers to Tunisia and to settle in what to her was virtually an unknown country by mid-September 1888.[45] Angiolina only authorized the sending of documents relating to the estate after further lengthy negotiations that continued for another month and a half.[46] In compensation, she obtained what she had asked for: she received ninety thousand French francs from the Tunisian administration, at least ten thousand more than Eva.[47] It was only at this stage that Angiolina also agreed to travel to Tunis.

In Tunis, the trajectories of Eva and Angiolina diverged even more. Santillana and the French consul in Florence firmly asked the Tunisian and French administration to assist Eva. But the French authorities did not trust her, nor did they support her. They knew that she had no financial resources to defend her case in the Tunisian courts, either in the Muslim court (*majlis al-shar'ī*) or the French court that had been established in 1883.[48] They also requested an inquiry into her daughter's nationality to establish whether Emma/Amina was a German citizen.[49]

Eva realized that she was on her own. Her husband was obliged to return to North America to go back to work.[50] Eva could not even count on the presence of her daughter. On leaving Florence, she stopped off in Marseille to place Emma/Amina in a convent, fearing the child might be kidnapped in Tunis.[51] Although Eva had followed her parents to colonial Algeria during her childhood, she did not draw on this prior experience to find a place in Tunisian society.[52] She did not reach out to the family of Emma/Amina's father, al-Sadiq Mu'addib, who had passed away four months before Husayn, at the end of February 1887,[53] nor did she bring her daughter to the Bu Hajib family: Eva rejected the very idea that her daughter Emma/Amina could be educated as a "good Muslim" in Tunis.[54]

Upon arriving in Tunis, Angiolina Fortunata Bertucci made a very different choice. She did not isolate herself, opting instead to marry 'Amr Bu Hajib and to become part of this Tunisian family. She settled down in the port town of La Goulette, near Tunis, and raised her daughter in the Bu Hajib's household.[55]

An Italian Woman Within a Muslim Household

At this point in these two women's journeys, how can we account for Angiolina Bertucci's resolve and ability to achieve her goals while Eva Keusch failed to win the claims she put forward for her daughter Emma/Amina? A comparison of the two cases clearly shows that the support of the French diplomatic and colonial authorities mattered less than overcoming social alienation by becoming acquainted with local languages and social codes and, more importantly, by finding a place in an influential family and household, such as that of the Bu Hajibs. In Tuscany, and even more so in Tunisia, Eva did not know in whom to confide. Angiolina, in contrast, could count on her better knowledge of Tuscan Italian and probably on a clearer apprehension of the judicial landscape and Tuscan society. More significantly, while settling in Tunis, Angiolina not only threw in her lot with key protagonists in Husayn's estate, but also, as a consequence of her new position within the Bu Hajib household, benefited from the social connections and legal knowledge of the head of the family, Salim Bu Hajib, who was then teaching at the prestigious Zaytuna Great Mosque. She could also rely on the administrative acquaintances of her husband's brother, Khalil Bu Hajib, who was working among the translators of the Tunisian administration.[56] Even before moving to Tunis, letters translated into Arabic on her behalf, probably by the Bu Hajibs, were sent to Santillana to defend her claims and her daughter's rights.[57]

In both Tuscany and Tunisia, societies that were still extremely patriarchal, this major difference between Eva and Angiolina's approaches led to significant consequences. Bertucci's integration into the Bu Hajib household reminds us that in such societies, households and families were still major social actors and spaces where subaltern women such as Angiolina could hope to gain or maintain a certain social position. The greater ease with which Angiolina and ʿAmr Bu Hajib assimilated and educated Maria/Myriam proves the capacity of Muslim and North African lineages and households to integrate outsiders as long as the integrated women and children did not threaten the status quo and the structure of the family, a topic better studied for the early modern period than for the colonial era.[58]

Indeed, the entry of European women into North African, and especially upper-class Tunisian, households was not new. At times, they even constituted a mark of high social status for prominent families.[59] Throughout the early modern period, from the sixteenth century to the second half of the nineteenth century, most of these Italian, Georgian, Greek, and Caucasian wives and concubines were slaves who had to convert to Islam. During the colonial period, European women entering Muslim households could still convert to Islam; this was probably the case for Angiolina when she married ʿAmr Bu Hajib. But these women were no longer enslaved, and they would rarely be treated as concubines. They would more often be married to the masters and sons of these households.

As the colonial administration imposed restrictions on the relationships between European settlers and colonized Muslims, these mixed marriages became scarce: they constituted fewer than 0.3 percent of the marriages in Tunisia by the end of the colonial period in 1959, and in Algeria, from the 1940s to the 1950s, only seventy-five Algerian Muslims married Europeans.[60] However, as rare as they might have been, women like Angiolina who married Muslim men complicated the binary division between colonized and colonizers.[61] Their position was ambiguous: they were neither colonized women nor the wives of European settlers, torn between their capacity to establish contact with local women and the temptation to erect barriers between their households and the surrounding society.[62] At the same time, women married to Muslim men were like any other women: they had to fight against and adapt to male domination, whether within their families or within a patriarchal system confirmed and supported by the French colonial administration.[63]

The Bu Hajibs Versus the Tunisian Sovereign

Even before becoming close with Angiolina, Shaykh Salim Bu Hajib and his sons were intimately involved in the Husayn estate as administrators and beneficiaries of his lands in Halq al-Wadi. Prior to his death, General Husayn "established [the properties there] as *awqāf*" to the benefit of Bu Hajib's sons as a group of private individuals (*ahlī*).[64] Husayn stated that he had "to bequeath something" to this family to thank Shaykh Bu Hahib for his loyal service and his sons for "their kindness towards him."[65] After the deaths of this generation of the Bu Hajibs, the income generated by the *awqāf* would then revert to a public cause (*khayrī*)—namely, supporting invalid soldiers from the Ottoman armies who had been "entrusted with defending the interests of Islam" and who would be housed in "the special establishment for military invalids."[66] At that stage, the task of managing the Halq al-Wadi properties would be transferred to the "*qāḍī* of the town where the beneficiary soldiers resided."[67] Husayn wanted to pay tribute to Muslim soldiers of the Ottoman Empire because while he had attained the rank of general, "he never actually fought for a holy cause or for the homeland."[68]

Unlike 'Ali Bey or Angiolina, the Bu Hajibs did not intervene in the conflicts surrounding Husayn's estate through secretive channels. Instead, they had to defend their rights publicly, in Muslim courts. The Tunisian governor initially acknowledged Bu Hajib's role in managing the Halq al-Wadi *awqāf*. However, 'Ali Bey quickly recanted his consent to the arrangement.[69] He and his entourage contested the deed establishing the *awqāf*, asking Muslim judges to examine its legality. From then on, for a two-year period running from 1887 to 1889, two judicial authorities dealt with this case: Muslim judges heard the dispute between Bu Hajib's lineage and the Muslim sovereign, while the French colonial court in Tunis (*tribunal de première instance*) was involved when non-Muslim European—mainly Italian—tenants (*inzāl*ists) of the Halq al-Wadi *awqāf* presented grievances or, more precisely, when they claimed that they did not know whether they should pay rent to the Bu Hajib family or to 'Ali Bey.[70] French colonial administrators asserted that they did not want to intervene, deeming the litigation a private conflict.[71]

Two years after the Tunisian justice acknowledged the Bu Hajibs' rights over the Halq al-Wadi *awqāf*, in 1891, the eldest son of the family, 'Amr, felt compelled to respond in an essay in Arabic to defamatory pamphlets probably written by Elmilik who, at the time, claimed that he was defending 'Ali Bey's interests.[72] It is not obvious whether 'Amr Bu Hajib addressed this manuscript to a specific audience or to a colonial readership, but a copy of his response has been kept in the Tunisian State Archives. Although I did not find the pamphlets penned by Elmilik during this particular conflict, 'Amr Bu Hajib's essay reveals how the men of his family acted to defend their rights against Elmilik's accusations.

In his response, 'Amr acknowledged that the deed establishing the Halq al-Wadi *awqāf* was quite unusual. The two original copies of the deed were not written in Arabic, as was often the case, but mostly in French, and unlike standard *awqāf* deeds, the two original copies signed by Husayn did not specify which family generations (*al-ṭabaqāt*) were involved or their succession (*ta'āqub*).[73] To explain such an unusual situation, 'Amr Bu Hajib argued that by the end of his life Husayn could no longer write, and because he could not rely on an Arabic-speaking secretary at the time, he dictated the contents of the deed to an unidentified individual who wrote them down in French. According to 'Amr, Husayn nevertheless certified the two copies of the deed by adding the last sentence and his signature in Arabic.[74] 'Amr added that as soon as they could, the Bu Hajib had had two trustworthy men translate the document from French into Arabic and three renowned Islamic notaries (or professional witnesses, *'udūl*) certify the deed.[75] One original copy was mentioned in a notarial register in Tunis; the other was sent to a judge (*qāḍī*) in Istanbul.[76]

The Bu Hajibs collected other legal decisions that supported their rights over the Halq al-Wadi *awqāf* from Muslim judges and from the most significant Islamic legal authority in Tunis, the *shaykh al-islām*.[77] At that point, they only needed a final approbation from a judge belonging to the Maliki school of Islamic jurisprudence, which was predominant throughout North Africa, not just in Tunisia.[78] But 'Ali Bey did not allow that and slowed down the process to such an extent that the Bu Hajibs asked Tunisian prime minister Bu 'Attur to intervene and queried the French *résident général* if they might settle the dispute with 'Ali Bey out of court.[79] Eventually, after some hemming and hawing, the Bu Hajibs won their case. Under pressure from the French colonial authorities, the *bey* was obliged to acknowledge the validity of their *awqāf* together with their rights to the properties in Halq al-Wadi.[80] This led ʿAmr Bu Hajib to conclude, in his reply to Elmilik, that the religious council (*majlis al-sharʿī*) of Islamic judges had considered the party representing the *bey*'s interests "like any other defendant" ("ḥakama ʿalā janāb al-bāy ka sa'ir al-khuṣūm"), adding that "the council has recognized the right of the party to whom this right falls." He concluded, "We are glad that such an event should occur in Tunis."[81]

The relationship between religious scholars and political authorities (including the Tunisian *beys*) is a classic trope in the literature on North Africa, and in Ottoman historiography more generally.[82] Study of the *ʿulamā'* is often associated with issues of secularization and Islamic reform.[83] ʿAmr's response to Elmilik highlights two interesting types of action that shed some light on these transformations. First, the Bu Hajibs could not ignore the increasing number of documents on Islamic legal issues written in European languages. ʿAmr penned a response in Arabic to accusations that were probably written in French, and in his response he acknowledged that the *awqāf* deeds were initially drawn up in French before being translated into Arabic. This was a sign of the times: Tunisian Muslim scholars could no longer live in a closed world and rely primarily on their command of Arabic prose and linguistic skills. If they wanted to be heard in public and legal debates regarding the press and the colonial administration, they needed to acquire a command of the French language in addition to their knowledge of Islamic legal institutions. Amr's brother Khalil acquired this sort of linguistic knowledge, becoming one of the translators working for the Tunisian administration by the end of the 1880s.[84]

Second, and more important, the Bu Hajibs used the new colonial context to assert their rights. ʿAmr Bu Hajib's response shows an awareness that the protectorate had ushered in a new balance of power in the relationship between Tunisian subjects and their sovereign. Under French colonial domination, ʿAli Bey was formally constrained to respect precedent and judicial hierarchies. Unlike his predecessors, ʿAli Bey could not abuse his authority by summarily dismissing a judge or confiscating assets of disputed ownership. In this context, the Bu Hajibs never openly opposed ʿAli Bey—they simply requested an audience. But by getting ʿAli Bey to compromise, the Bu Hajibs contributed to refashioning Tunisian citizens' relationships with their sovereign. In ‘Amr Bu Hajib's response, ‘Ali Bey was presented not as a final resort or fount of justice, but as subject to the law like any other Tunisian national. In this sense, the Bu Hajib's actions helped Islamic justice acquire greater autonomy from the Tunisian *bey*. They thus played a role in a broader process of secularization and Islamic reform.

Elmilik, the Jewish Courtier Who Challenged Tunisian and French Authorities

Deeply involved in the conflicts between ‘Ali Bey and the Bu Hajibs, maneuvering to represent the Tunisian sovereign in Tuscany, the last character of this chapter, Léon Elmilik, was central to the Husayn estate affair. At the very start of the affair, Elmilik claimed that he was Husayn's primary creditor. Certain of Husayn's debts, such as those to his physician, pharmacist, and servants, were paid very quickly, and those creditors were not heard from again.[85] But up until 1889, Elmilik, as well as Eugène Cohen Rosa, who had worked for Husayn during the Scemama affair before leaving his service in the late 1870s, continued to make significant claims against Husayn's estate.

While Rosa was very discreet and is barely mentioned in the primary sources, Elmilik did not shy away from scandal, and even courted it. He felt that Husayn had not sufficiently recompensed him for services rendered. He claimed that Husayn owed him Fr.800,000.[86] Extremely active and particularly prone to quibbling, Elmilik firmly defended his right to attend the inventory of Husayn's papers.[87] In winter 1887, as seen in the previous chapter, he went so far as to sue the French consul—his own representative, as he had French citizenship at that point—who would not allow him to attend the inventory, triggering the intervention of Italian policemen in the French consulate and resulting in a very serious diplomatic incident between France and Italy.[88] The following year, Elmilik was involved in no less than three legal cases in Italy and Tunisia: in addition to his extensive involvement in many conflicts over Husayn's estate, he decided to sue one of Caïd Nessim Scemama's relatives and another man who had showered insults on Elmilik during a meeting in Halq al-Wadi.[89] Last but not least, up until the summer 1889, Elmilik prevented the French consulate from sending Husayn's papers from Florence to Tunis, being very much aware of how firmly it was in his own legal interests to keep the Husayn estate affair in Tuscany.[90]

Elmilik took his accusations against the French authorities very far, to the point that he clearly broke article 9 of the French-Italian diplomatic agreement which prevented him, as a French citizen, from "invoking local justice against his consul."[91] As a result, after the Florence incident, Santillana, the French consul, and Ghanem all repeatedly expressed their desire that Elmilik be made an example of, either by having him arrested or by expelling him from Italy and Tunisia.[92] But, astoundingly, Elmilik was neither expelled nor stripped of his French nationality. The French authorities never imposed sanctions, fearing that this "would certainly fuel further press comments."[93] And while the courts in Florence found Elmilik's claims against Husayn unwarranted, the French authorities gradually yielded to each of Elmilik's demands. Not only did they authorize him to attend the inventory of the estate in the French consulate, but Elmilik also obtained copies of inventoried items. In late June 1888, the agreement with ʻAli Bey granting 5 percent of the revenue of Husayn's estate to Elmilik was reinstated, against the colonial authorities wishes.[94] Finally, in 1889, the secretary general to the Tunisian government granted Elmilik an indemnity of Fr.79,608 (of the Fr.200,000 that Elmilik had initially sought) once Elmilik allowed French authorities to send Husayn's papers to Tunis.[95]

Elmilik's fierce determination, his eagerness to sue each and every opponent of his claims (including the French authorities), and his impressive ability to wrest financial compensation from unwilling parties are very specific features of his character. In this sense, his case is not at all representative of the history of North African Jews in the Mediterranean. Nonetheless, despite its specificity, the case of Elmilik provides us with avenues to analyze the agency of North African Jews who were no longer under Muslim "protection" and had gained French citizenship in the colonial context.

How Does Elmilik's Story Fit Into the History of North African Jews?

Emilik's case does not correspond to "much of the literature on the Jews of the Maghreb" that, according to historian Daniel Schroeter, has long subscribed to the history of "singular [Moroccan, Algerian, Tunisian, or Libyan] Jewish communities" according to the political divisions of the North African states.[96] Elmilik, who hailed from Algeria and as such was of French nationality, never ceased interacting with Jewish communities in

Tunisia, Algeria, and Tuscany, as well as with Muslim dignitaries and their entourages in Italy during his missions for Husayn.[97] His story spanning all these countries reminds us that this mediator and other North African Jews who circulated between the two rims of the Mediterranean Sea were part of what Schroeter has coined a "trans-regional network of connected Jewish communities."[98]

Elmilik's case does not help answer a question that has long been predominant in scholarship on North African Jews. His story is not about how Jews were treated in the Mediterranean, nor does it convey a rosy picture or a "lachrymose conception" of the relations between Jews and Muslims.[99] Elmilik does not conform at all to a view that has dominated that same literature: the idea that during the colonial period, French North African Jews distanced themselves from Muslims and became "loyal" and "grateful" to the French republic for bringing "progress" and "emancipation" to the Jewish communities in North Africa.[100] As we saw, Elmilik repeatedly disobeyed French authorities, confronting and challenging them as much as he could. Whether through the publication of accusatory pamphlets against Husayn and then against the Bu Hajib family or by launching legal cases against the French on several fronts, Elmilik turned all the legal tools at his disposal and his knowledge of the Italian, French, and Tunisian legal systems back against his adversaries—including the French authorities.[101]

Elmilik might be seen as atypical, even exceptional, but building on similar trajectories, historians of North African Jewries have also nuanced the idea of a complete Jewish loyalty to the French republic and the "French republican claims of Jewish betterment through the nation-state's assimilatory embrace."[102] For instance, historian Joshua Schreier's study of the Moroccan-born Jewish merchant Jacob Lasry (1793–1869), who settled in the Algerian city of Oran, demonstrated that it was the French "who created the singular [Jewish] community that they later boasted of . . . having emancipated."[103] In other words, in Oran and other regions of colonial North Africa, the "community of *Israélites indigènes* did not form a preexisting and unified social group; rather, it was a product of the dynamics of French imperial expansion."[104] More broadly, historians have demonstrated how European Jewish scholars, on the one hand, and institutions such as the French Alliance Israëlite Universelle, on the other, "orientalized" Mahrebi Jewries, portraying these communities as backward in order to "modernize" them and rescue them from their traditions.[105]

Elmilik, as an exceptional case, does not tell us much about the ambitions of European Jewish organizations such as the Alliance Israëlite Universelle or France's endeavors to unify North African Jewries under their aegis. Rather, it draws our attention to specific individuals who were simultaneously using and even wagering the social credit they had acquired (in the form of social relations and legal statuses) to make legal and financial claims, to reach new thresholds of negotiation, and to test their opponents' limits.[106] One wonders who among North African citizens was able to afford such risks. Who could sue people and institutions with reckless abandon? Who would not hesitate before attacking their national tutelary power? 'Ali Bey, the Bu Hajib family, Eva Keusch, and Angiolina Bertucci were far more cautious. Was this risky attitude specific to people like Elmilik who had once belonged to a subjugated religious minority, been granted the political rights of French citizens, and learned a great deal from legal experiences outside of colonial North Africa, in Mediterranean venues such as Tuscany? In this sense, studying the agency of a North African Jew such as Emilik opens up new questions about North African colonized subjects as a whole.[107]

French Colonial Domination and Other Historical Dimensions of Modern North Africa

Each of the individuals involved in the conflicts over Husayn's estate acted in specific ways and developed multiple, interesting relationships with French colonial authorities depending on their social class, their gender, or their belonging to religious communities and households. Under high scrutiny by the colonial administration, 'Ali Bey managed to bypass French surveillance and act secretly, relying on his family, former officers, and intermediaries of the precolonial Tunisian government. By allying with the Bu Hajib household, the Italian Angiolina Bertucci became a more effective negotiator than the French-German Eva Keusch, who did not succeed in gaining the French officers' trust and backing. 'Amr Bu Hajib had to take into account the pervasiveness of French documents and French language within the Tunisian legal institutions, but also took advantage of the new colonial balance of power to get 'Ali Bey to compromise. And, although a French citizen from Algeria, Léon Elmilik did not hesitate to cause a scandal and sue the French representative in Italy so as to catch the attention of every character in this story.

In each case, French colonial domination constrained and influenced these individuals' abilities to act. At the same time, 'Ali Bey, Angiolina Bertucci, and Léon Elmilik all succeeded in challenging France's power to oversee and repress individuals and authorities in the Mediterranean basin. To do so, they relied on households, social connections, and Tuscan courts, spaces and institutions that were already instrumental during the precolonial period. Moreover, most of these players did not base their actions on the French colonial agenda, whether they were for it or against it. Instead, litigation about land, credit, or Islamic values (in the case of the Bu Hajib) triggered most of their actions.

Studying these varied sorts of agency allows us to recover dimensions that are often overlooked in the history of colonial Tunisia, such as the lasting role of Muslim households, the clandestine actions of the Tunisian sovereigns, and the importance of Tuscany and the Italian courts as venues for North Africans to challenge French authorities at the end of the nineteenth century. How can we write a history that takes into account colonial domination and these other dimensions of history at the same time? Were these other forms of action, and the histories that they point to, alternatives to the colonial world—as subterranean domains eluding the attention of colonial authorities?[108] Or do these actions reveal instead aspects of history that enveloped colonial domination?[109]

Our attempt to look closely at various forms of agency is not meant to fragment the history of these societies into diverging accounts, in a "limitless and arbitrary proliferation of interpretations."[110] Rather, it aims to problematize the central position of colonialism in the study of the modern history of a particular North African region in order to save from oblivion "other signifying systems which had lost the battle for legitimacy and have thus been 'forgotten.'"[111] The various characters in this chapter might be seen as exceptional, but each nuances and enriches our understanding of this period. 'Ali Bey is a reminder of the various layers of political action in colonial Tunisia. Elmilik showcases that even while becoming French, North African Jews could still challenge colonial authorities. Finally, Angiolina Bertucci provides an illuminating instance of a European woman who found a place in a Muslim household during the colonial period and challenged the binary division between colonized and colonizers.

SIX

Husayn's Legacies in Colonial Tunisia

An Epilogue

AFTER SO many years of litigation, which of the claimants had won? And more important, to what extent did the Husayn affair contribute to the transformation of legal norms and archival practices in colonial Tunisia?

Diplomatic conflicts between Tunisian, French, and Ottoman authorities—and, to a lesser extent, Italian authorities as well—over Husayn's estate came to a close in early 1888; the private disputes over his assets ended in the early 1890s. At the international level, as we will see in the first section of this final chapter, the French authorities can be seen as the primary diplomatic winners in this story. First, the French consul in Florence managed to take control of Husayn's estate and private papers, dispatching strategic files to the French diplomatic archives in Paris (Quai d'Orsay) and depositing a sizable number of documents in both the state and colonial archives in Tunis. Second, when the remaining claimants (Angiolina Bertucci, Eva Keusch, Léon Elmilik, and Eugène Cohen Rosa) agreed to send Husayn's papers to Tunis, they also agreed to settle in Tunis, therefore coming under Tunisian and French legal and political authority. In this sense, France won yet another battle. Italian courts would no longer make decisions related to Husayn's estate. From then on, French and Tunisian judges would be in charge, within a judicial hierarchy that was more and more under the French colonial aegis.[1]

However, French victory in the Husayn affair was not complete. In order to assert their control over Husayn's papers, parts of his estate, and the remaining litigants in this affair, the French had no choice but to make use of, and therefore maintain, provincial Ottoman political practices related to slavery, such as contractual clientage (*walāʾ*) over manumitted slaves, as well as the procedures for negotiating with claimants and creditors over this type of estate. In a sense, then, the Husayn affair had an ambivalent effect on the French colonial presence in Tunisia: on the one hand, it reinforced France's grip on the Tunisian state archives and legal system; on the other hand, it revealed the lasting effects of Ottoman political culture in Tunis and contributed to a dual process of "Ottomanization" and "localization" of the French colonial system, to borrow the expressions coined by historian Ehud Toledano.[2]

The most successful individual claimants to Husayn's estate were not those who had initially sided with France, ʿAli Bey and Eva Keusch, mother of Emma/Amina. The winners were Angiolina Bertucci and the Bu Hajib household, who had both been part of Husayn's daily life. As we saw in the previous chapter, Angiolina received Fr.90,000 from the Tunisian government, the Bu Hajib sons asserted their rights over Husayn's *awqāf* in Halq al-Wadi, and Angiolina and her daughter Maria/Myriam became fully integrated into the Bu Hajib household following Angiolina's marriage to ʿAmr Bu Hajib.

In the second section of this chapter, I follow the transformations of the Bu Hajib household from 1890 to the 1920s to elucidate two major aspects of the history of colonial Tunisia that have been underappreciated in scholarship. I emphasize the lasting importance of Tunisian elite households, which, following historian Jane Hathaway, I define here as a kinship group and those who benefited from the patronage of the household head "whether or not these [were] related to him."[3] Historian Leslie Peirce has stressed the importance of the household in the early modern Ottoman world, characterizing this structure as "the fundamental unit of political as well as social organization" and "the matrix for a complex network of relationships that transcended the public and the private, the personal and the institutional."[4] Households as crucial places for political and social action and as hubs that still connected Tunisian elites with Middle Eastern and Ottoman elites have been underinvestigated in the historiography of colonial North Africa.[5] Here, I contend that the Bu Hajib household maintained strong relationships with two centers of the Ottoman Empire (Istanbul and Cairo) through two women in particular: Maria/Myriam, who followed her husband, Tunisian reformist Ali Bash Hanba, to Istanbul during the First

World War; and Khalil Bu Hajib's wife, Princess Nazli Fadil, who was the niece of Khedive Isma'il, the ruler of Egypt and Sudan from 1863 to 1879.

Although the Bu Hajib sons worked with the colonial administration and Khalil Bu Hajib later rose to become prime minister (1926–32), the family's interactions with the French should not be interpreted solely through the classical frameworks of collaboration/resistance. Rather, Shaykh Salim Bu Hajib's and his sons' interactions with French authorities should be understood as part of a broader competition between Muslim scholars and statesmen within Tunisian society. French authorities were only one player among many in these social rivalries.

Therefore, the entirety of this chapter seeks to nuance the now widespread idea that the resolution of legal cases such as Husayn's within a colonial context only strengthened colonial domination and contributed to the shaping of a colonial legal hierarchy. To make their point, French authorities had to rely on a number of local legal tools and practices. Moreover, they were not the sole winners in this legal battle, since a Muslim household also benefited greatly from Husayn's estate. And while Shaykh Salim's sons had to accommodate themselves to the French domination of Tunisia, they also managed to maintain their social and political connections with the Ottoman Levant.

The Victories of the French Colonial System

From Husayn's death in 1887 until February 1889, French authorities in Florence faced a string of obstacles when it came to collecting, translating, and then archiving the vast number of papers related to Husayn's estate. For many months, claimants to the estate prevented the papers from being sent from Florence to Tunis. Furthermore, the French authorities did not have the financial or human resources to translate most of Husayn's documentation from Arabic into French or to catalog every individual document.[6] The attorney, David Santillana, was proficient in Arabic but had little time for such tasks, mainly producing oral translations as the documents were being inventoried.[7] The official interpreter, Chekri Ghanem, was at a complete loss "amidst the cries and remonstrations" at the consular chancellery, caught between his multiple roles as interpreter, copyist, and archivist and the contradictory orders he received from French consul Louis de Laigue and Santillana.[8]

In February 1889, a year and a half after Husayn's death, the French authorities attained their goal: Husayn's papers left the shores of Tuscany. The documents were shipped in three huge boxes "bearing the consular seal." The boxes contained "five trunks also under seal," with "seven travel bags or small cases" in each trunk.[9] The papers were then transferred to two separate places: legal documents were shipped to the clerk's office at the Court of First Instance in Tunis, where they would be used in the continuing litigation involving Elmilik, among others, while other papers deemed more sensitive (because they concerned the French authorities in Tunis) were dispatched to the French Ministry of Foreign Affairs in Paris.

The French authorities thus came into possession of precious documents, some of which Husayn had concealed from the Tunisian state and administration for more than a decade. From the inventory of the papers in Florence to their archiving in Paris and Tunis, French authorities kept on reworking the former logic of these documents and therefore their original rationale. Husayn's books and papers were selected for their presumed interest to the French authorities and distributed accordingly. For example, French consular authorities sent the Arabic books from Husayn's private library in Florence to the colonial administration in Tunis without even inventorying them; they only listed the ones in European languages.[10] French diplomats paid closer attention to writings they considered "political"—notably, Husayn's correspondence with former prime minister Khayr al-Din and European and Ottoman political figures, together with the legal statement that Husayn had penned in the late 1870s against former prime minister Mustafa b. Isma'il, who had acquiesced to France's colonization of Tunisia. Some of these "political" documents were "sent straight to the Tunisian government," but the more sensitive ones are still held at the new location of the French diplomatic archives in the Parisian suburb of La Courneuve.[11]

The inventorying, selection, and transfer of papers to Tunis and Paris appeared to mark a French victory over Italy and the Ottoman Empire, which had to acknowledge that the French administrative and diplomatic hierarchies were entitled to oversee the affairs of Tunisian subjects. The French administration rewarded all the intermediaries and civil servants who had worked to further French imperial interests, including Ghanem and Santillana, who was granted a prestigious French medal, the Knight's cross (*croix de chevalier*).[12]

Transferring Husayn's papers from Italy to France and Tunisia resulted in yet another French victory: the remaining litigants in the Husayn affair (mothers Eva and Angiolina and creditors Elmilik and Rosa) were no longer able to act from Tuscany and use Italian justice.[13] With most of the documents related to Husayn's estate now in the clerk's office at the French Court of First Instance, the claimants had to go to Tunis if they wanted to solve their legal issues and continue to press claims to the disputed legacy.[14] In Tunis, the entire judiciary, both Muslim judges and especially the French courts established in 1883, were under French colonial control. After the arrival of the documents, from 1888 onward, a joint real estate tribunal was set up, theoretically dependent upon the Tunisian government but in fact presided over by a French expert.[15] The Court of First Instance, for example, scrutinized 'Ali Bey's claims to Husayn's *awqāf* in Halq al-Wadi. In 1889–90, Giuseppe de Cardoso claimed that the Tunisian sovereign had to reimburse the Fr.50,000 he had lent to Husayn; the case went as far as the Court of Appeals in the neighboring colony of Algeria.[16]

The resolution of the disputes over Husayn's estate thus confirms the conclusion of many works on legal pluralism: as with similar legal disputes, Husayn's case reinforced the colonial hold over local legal order. In a country with more than one judicial hierarchy, where litigants could present their claims to both Muslim and colonial courts, disputes tended to spawn further disputes.[17] The increasing number of disputes led the colonial authorities to intervene and hierarchize the various judicial bodies involved.[18]

Perpetuating Ottoman and Local Practices

If we consider the legal and archival outcomes of this affair, there can be little doubt that the French authorities were victorious on many fronts. I argue, however, that this French victory was incomplete for several reasons. First, French officials were not the only ones who benefited from the legal experiences gained from the Husayn affair. Over the course of the disputes, Santillana augmented his knowledge of Islamic law by a considerable measure. The Husayn affair was one of many cases that led to his being recognized as a renowned expert on Islamic law. More than two decades after the resolution of Husayn's estate, Santillana's expertise became crucial for the Italian state, which was developing into a colonizing force in

North Africa. After conquering the Ottoman province of Tripoli in 1911, the Italians consulted Santillana on questions relating to Islamic law. A few years later, in 1919, in order to support Italian administrators in Libya, Santillana published a translation of al-Khalil's *Mukhtaṣar*, a concise handbook from the "foremost source of the Mālikī school" of Islamic law that treated various social and economic actions in daily life, such as contracts, credits, and sales.[19] He also wrote a major summary of Islamic law according to the Maliki interpretation; the first volume, titled *Istituzioni di diritto musulmano malichita con riguardo anche al sistema sciafiita*, was published in 1926, and the second posthumously in 1938.[20]

More important, in order to gain access to Husayn's papers and to control all remaining litigation over his estate, French authorities were forced to perpetuate practices from Ottoman times which, though fundamental to Tunisian society, ran counter to French political and legal practices. For instance, Santillana may well have advocated for the abolition of contractual clientage (*walā'*) "by a committee of Muslim legal experts, as a logical consequence of the abolition of slavery" promulgated in Tunis in a series of *beylical* decrees between 1841 and 1846.[21] Yet the French and Tunisian authorities still relied on contractual clientage over manumitted slaves because it legitimized both 'Ali Bey's claims to Husayn's inheritance and some of their own strategic interests.

French colonial authorities desperately needed to control the strategically located palace that Husayn had rented to the French army since 1882. The building, located at one of the entrances to the medina of Tunis, had been transformed into barracks in the early 1880s, and the French officers housed there believed that their government had to acquire the palace to control that part of the city. Interestingly, they justified this belief on the grounds that they had been acting as a substitute force for Tunisian soldiers since the 1881 conquest.[22] The French army eventually purchased the building in 1890.[23] Husayn, like his mamluk predecessors, was just a cog in the transmission of this piece of real estate.

The French colonial authorities' pragmatic approach to slavery in Muslim lands is now well known.[24] Inès Mrad-Dali's work on "black Tunisians" from 1846 onward, for instance, has shown that colonial authorities were reluctant to end domestic slavery, claiming that they did not want to interfere in Muslim domestic affairs, despite knowing that African slaves were serving not only Muslim families but European settlers in Tunis as well.[25]

In Husayn's case, as for other mamluks who had served the Tunisian sovereign, the French could not get rid of *walā'* because it was interwoven with practices and ideas relating to property and state service. In other words, Husayn was not simply the manumitted slave of an important Tunisian. He was also the servant of the Tunisian state. As such, he managed state properties and was granted state lands. The gradual disappearance of slavery in colonial North Africa thus also needs to be viewed in light of the personal relationships of dependence and slavery, patrimonial strategies, and colonial interests.[26] Tunisian historian Béchir Yazidi has suggested that French colonial authorities partly replaced the Tunisian sovereign as the "main distributor of Tunisian land" up until the introduction of a system for registering property in Tunisia in 1885.[27] In the same capacity, because they sought to control 'Ali Bey's actions, one could argue that the French colonial authorities took charge of his former slaves and their estates as well.

In addition to maintaining *walā,'* the protectorate administration perpetuated other practices in its relations with its Muslim and European subjects. Throughout their management of the Scemama affair, French administrators repeatedly criticized Tunisian authorities, and Husayn in particular, for their tendency to negotiate with individuals over Scemama's estate. French administrators said they wanted to avoid such a precedent—along with the attendant string of disputes and dilapidated finances—by whatever means necessary.[28] Yet, just like the Tunisian agents in the Scemama affair, the French authorities brokered compromises between those with claims to Husayn's estate: Léon Elmilik, Eva Keusch, Angiolina Bertucci, and Eugène Cohen Rosa. For instance, the protectorate administration judged Elmilik's claims to Husayn's estate to be largely unfounded;[29] nonetheless, after arbitration with the *bey* of Tunis, Elmilik in 1891 received an indemnity of Fr.60,040, four times more than Rosa, who only received Fr.15,000.[30] Like the Tunisian authorities prior to the conquest, French authorities struggled to contain demands and disputes arising from this type of affair. Indeed, once the estate papers had been transferred to Tunis, Rosa threatened to take another case to court in late 1890.[31] And when Elmilik died in 1891, his sons Moïse and David, fed up with waiting for the outstanding sums owed to their father, also threatened to go to court.[32]

French victory was incomplete for one final reason: all the compromises struck by the French administration paradoxically resulted in weakening the cases of the two main litigants who had initially sided with France

('Ali Bey and Eva Keusch) while strengthening those of Husayn's intimates and close allies (Angiolina Bertucci and the Bu Hajib family). French colonial authorities did not succeed in defending 'Ali Bey's interests. Despite being Husayn's legal heir and theoretically under French protection, the Tunisian sovereign did not benefit from Husayn's estate—quite the contrary. Expenditures arising from the estate, after receipts had been deducted, were estimated at Fr.425,000 in 1890.[33] Two years later, 'Ali Bey had to make good on an estimated deficit of Fr.481,999.73.[34]

That same year, in 1892, more than three years after coming to an agreement with the *bey*, Eva Keusch and her daughter Amina/Emma, who had joined her in Tunis, found themselves weak, hungry, and isolated.[35] Eva was anemic:[36]

> On the 11th of this month when I came to the *résidence* to talk to you about my case, I had to lean against the wall so as not to collapse from weakness. I am ill and for the past four or five weeks, I have only eaten a little bread and water. In a word, I am dying of hunger at home. I beg you, please give orders for me to be paid, so that my state of health does not get worse, for it is a shame to die of hunger.[37]

The balance drawn up by the protectorate administration between 1890 and 1892 shows that the greatest beneficiary of the transactions surrounding Husayn's estate was, surprisingly, Angiolina Bertucci. In 1890, she received at least Fr.120,000. Such an outcome meant that the theoretical hierarchy of Husayn's heirs was overturned. Instead of obtaining one-sixth of Husayn's assets in accordance with his will, Angiolina got more than Husayn's legal heir, the *bey*, who was required to repay all Husayn's debts. She also received more than Eva, the mother of the other legatee. In addition to the Fr.90,000, Angiolina was able to garner another Fr.30,000 for the sale of a Fr.50,000 note that Husayn had signed over to her on October 1, 1866.[38]

As seen in chapter 5, Angiolina and her daughter Maria/Myriam joined the Bu Hajib household, which had successfully defended its rights to Husayn's *awqāf* in Halq al-Wadi. This leads us to the Bu Hajibs, the other main beneficiaries of the competition over Husayn's estate. Here we will attempt to understand what the Bu Hajibs and Angiolina made of their victory, how their household reshaped Husayn's material and political legacy during the colonial era, and how it stayed connected to the Muslim world and more precisely to the important Ottoman political centers of Cairo and Istanbul.

A Household Between French Domination and The Muslim World

The Late Nineteenth-Century Battles Over Islam

By the early 1890s, Shaykh Salim Bu Hajib's household (figure 6.1) reached the status of Tunisian notables. Once the *bey* acknowledged Bu Hajib's rights over the *waqf* in Halq al-Wadi, the Bu Hajib sons managed these properties

FIGURE 6.1 Shaykh Salim Bu Hajib and his sons 'Amr, Husayn, and Khalil (pictures Habib Bouhageb)

and collected the income they generated, exceeding 41,150 piasters on average between 1886 and 1890.[39] Their renown and influence also derived from the stature of their patriarch, Shaykh Salim, a respected teacher at the Zaytuna Great Mosque, who was involved in Islamic reform and who established new publications and Muslim educational institutions. The household's prosperity gave rise to envy: in 1890, seventy-five thousand Tunisian piasters were stolen from the Bu Hajib residence.[40]

Shaykh Salim's entire life of teaching and reform might appear contradictory if perceived solely in terms of resistance to or collaboration with French colonization. On the one hand, his former students attested to the crucial role he played in initiating reform in the Muslim world and in generating awareness of the importance of education.[41] On the other hand, the *'ulamā'* in Tunis, including the shaykh, did not really react to the 1881 conquest by the French. Additionally, at least two of his sons were educated in French schools, and Salim was a member of the Khalduniyya, a learned society founded in 1896 under the patronage of the French colonial administration.[42]

These various facets of Salim's life and teachings become less contradictory once we situate his trajectory alongside that of Husayn and other public figures in the province of Tunis from the 1860s onward, taking into account their common desire to reform both Islam and Islamic education. Bu Hajib's sons attended French schools during the shaykh's stay in Tuscany in the 1870s, when he became better acquainted with the Italian approach to teaching and began to admire certain key references from Western culture, such as Machiavelli.[43] Similarly, his commitment to his role as a teacher, both at Zaytuna and as a member of the Khalduniyya, placed him squarely within the tradition of great scholars from the Grand Mosque who sought to reform and adapt Islamic teaching to the "modern" era.[44] We find a third important indication of Salim's commitment to Islamic and educational reforms in his major role in hosting Egyptian theologian and judge Shaykh Muhammad 'Abduh during two stays in Tunis, from December 6, 1884, to January 4, 1885, and again in September 1903, two years prior to 'Abduh's death.

A disciple of Jamal al-Din al Afghani, another major Muslim scholar, 'Abduh advocated for the reform of Islam, the "renovation of the Arabic language," and the defense of the Muslim peoples.[45] Prior to 'Abduh's first visit in 1884, Salim Bu Hajib and Muhammad al-Sanusi corresponded with al-Afghani, who was collaborating with 'Abduh on a newspaper, *al-'Urwa al-Wuthqā*, in Paris.[46] The idea was that 'Abduh would travel from Paris to Tunis to solicit funds for the newspaper, which had initially been sponsored by Husayn. At that time, according to historian Arnold E. Green, 'Abduh presented *al-'Urwa al-Wuthqā* as "an anti-Western secret society allegedly founded by Jamal al-Din al-Afghani in Hyderabad and having numerous branches throughout the Islamic world," including in Tunisia.[47] During the second visit, in September 1903, Salim's son Khalil Bu Hajib hosted 'Abduh for two weeks in La Marsa, a well-to-do coastal town north of Tunis. On this occasion, Salim prepared a lavish dinner at his house in Halq al-Wadi for this prestigious guest.[48] 'Abduh's two visits to Tunis were only part of a lengthy exchange that the Egyptian had initiated with Tunisian *'ulamā'* and dignitaries about reforming Islam, education, and the press back in the 1870s.[49]

The desire to educate, to reform religious culture, and to adapt the circulation of knowledge in the Muslim world were far more fundamental for Shaykh Bu Hajib than his relationship with the French colonial authorities.[50] Like many Muslim reformists of the period, Salim accepted an ad hoc alliance with the French authorities in order to achieve the goals that had driven his course of action since the 1860s.[51] From this perspective, France was not pivotal. The French authorities were only one of many protagonists in a competition between Muslim elites supporting what they considered to be Islamic orthodoxy and those, like Salim, who were in favor of *iṣlāḥ*, or reforming Islam and its modes of government.

These two camps clashed on the occasion of Shaykh 'Abduh's second visit to Tunis. The first time 'Abduh came, the Bu Hajibs had extended him a warm welcome in their residence at Halq al-Wadi, and the *bey*s and other Tunisian personalities were similarly attentive.[52] But after his second visit more than two decades later, the supporters of "orthodoxy" were vociferously opposed to 'Abduh. They claimed that his visit "attacked Muslim orthodoxy and the worship due to the saints."[53] A Zaytuna teacher, Salih al-Sharif, berated 'Abduh for "condemning visits to the tombs of the saints and Prophet" which, according to him, was an integral part of Muslim devotion.[54] Muslim scholars who labeled themselves "conservatives" petitioned the Tunisian prime minister in 1904 requesting that he ban a monthly journal, *al-Manār*, founded in 1898 by one of the shaykh's disciple, Rashid Rida, to promote 'Abduh's views.

Maria/Myriam in Istanbul: Revolutionary Fervor

After 'Abduh's second visit, the Bu Hajib family changed considerably. Angiolina's daughter, Maria/Myriam, who had been brought up in the Bu Hajib household and converted to Islam under the name Farida, married a Tunisian political leader, 'Ali Bash Hanba, in 1908 when she was twenty-five.[55] Like the Bu Hajib sons, Bash Hanba had studied at the Sadiki College, a prestigious secondary school founded in 1875 by Khayr al-Din to train new officials for the Tunisian administration. Bash Hanba then graduated with a law degree from the Faculty of Paris and founded the *Tunisien*, the first French-language newspaper published by Tunisians between 1907 and 1911, with a section in Arabic after 1909.

The marriage between Farida and Bash Hanba enabled the Bu Hajibs to consolidate their close political and social ties with one of the leading figures of the Young Tunisian reformist movement that emerged in the late nineteenth and early twentieth centuries. The Young Tunisians were a small group of educated Tunisians often seen as precursors to the Tunisian nationalist movement. Their main goal was to improve the social and political rights of their fellow citizens by reforming and liberalizing the French protectorate regime. In 1911, as summarized by historian Kenneth Perkins, "an Italian tram driver ran down a Tunisian child" in the context of the Turkish-Italian war in the neighboring province of Tripoli. Young Tunisian leaders Bash Hanba and his comrades were arrested and exiled after they launched a boycott of the trams in reaction to the death of a child and after they demanded the election of a Tunisian representative to a consultative institution.[56] Bash Hanba stayed with his brother in France for a while, still seeking a political compromise that would allow him to return to his native country.[57] But in 1912, having failed to convince the French bureaucrats to compromise, he settled in Istanbul; Farida soon followed.[58]

In Istanbul, 'Ali and Farida joined more than three hundred Tunisians, as well as Algerians and Tripolitans who had come to Istanbul in large numbers in the wake of the 1908 Young Turk Revolution that had led to the restoration of the 1876 Ottoman constitution and the organization of parliamentary elections.[59] Word had gotten out that "the Turkish government bestow[ed] significant advantages on emigrants" from North Africa. A secret committee operated in Tunis to encourage the *hijra*, or holy emigration, to the former imperial capital of "Muslims [who had] sunk" into poverty "under the dominion of Christian powers."[60] In 1910, spectacular processions of *muhājirūn* (exiles who had accomplished the *hijra*) paraded in the streets of Istanbul.[61]

The Italian invasion of Tripoli in 1911, and the eventual defeat of the sultan's troops in Libya, deeply traumatized both the Ottoman and colonized North Africa populations. Against the backdrop of the war, in which Italy and the Ottoman Empire clashed over the Ottoman province of Tripoli, Bash Hanba founded a Muslim Union Movement (*al-ittihād al-islāmī*).[62] He relinquished his Tunisian citizenship and adopted Ottoman nationality in Istanbul.[63] There, Bash Hanba worked for the Ministry of Justice and in 1914 joined the legislative section of the Council of State.[64] The following year, in May 1915, Minister of War Enver Pacha appointed him head of the Special Organization (*Teşkilat-ı Mahsusa*).[65]

The Special Organization (SO), also known during the First World War as the Organization for Oriental Affairs (*Umur-i Şarkiye Dairesi*), was in charge of several official missions: settling refugees in Istanbul, conducting propaganda, and overseeing prisoners of war by putting them to work or assigning them to other fronts.[66] In his position as SO head, Bash Hanba incited forces to rebel against Allied troops of the Entente (initially France, the United Kingdom, and Russia) across a broad swathe of territory running from Morocco to the East Indies.[67] Above all, under Bash Hanba, the SO was involved in combating all separatist movements. Historians have shown that it was this organization—headed by Bash Hanba and operating under oversight by the Committee of Union Progress members—that organized and perpetrated the genocide of the Armenian population.[68]

Bash Hanba died of a fever on October 29, 1918 and was buried in Istanbul near the Çırağan Palace.[69] He left behind a daughter raised by Farida.[70] Soon after her husband's death, Farida returned to Tunis. This was not her first return to Tunis: a few years earlier, she had come back for a final visit with her mother Angiolina, who passed away on May 28, 1914.[71] The final decade of Farida's and 'Ali Bash Hanba's life together is fascinating because it brings to light unexpected aspects of both the transformations of the Bu Hajib household and the history of burgeoning Tunisian nationalism. Farida's exile in Istanbul is clear evidence of the lasting connections between the Bu Hajib household and the center of the Ottoman Empire. Her visit to her mother in Tunis also points to the fact that Angiolina and Farida found stable social positions among the Tunisian elites from the late 1880s to the 1910s.

More broadly, Farida's husband's political activities in Istanbul prior to the First World War complicate the commonly held idea in the historiography of colonial Tunisia and Algeria that nationalism in both countries was born after the war and the political failure of reformists like the Young Tunisians and the Young Algerians. 'Ali Bash Hanba's violent policies directed against non-Muslims—along with the commitment of other North African activists to defending Muslim populations—also qualify the idea that North African nationalism and separatism were born in a social milieu far from Istanbul, among North African students and workers educated and politicized in French factories by the French Communist Party and European trade unions.[72]

In the aftermath of the Italian colonization of Libya, amid the First World War, the Young Tunisians and Young Algerians had already rephrased their reformist slogan in Istanbul and Europe into a nationalist and separatist claim. This advocacy for radical change spread across the center of the Ottoman Empire, North Africa, and Europe in the 1910s and 1920s. For instance, in Berlin in 1916, Salah al-Sharif, an exiled Algerian who had taught in Tunis at the Zaytuna Great Mosque and interacted with 'Ali Bash Hanba in Istanbul, "gathered a group of North African exiles to found" the Committee for the Independence of Tunisia and Algeria amid the tumult of arms and executions.[73] That same year, in Geneva, Ali's brother Muhammad Bash Hanba—who frequently traveled to Istanbul—founded *La Revue du Maghreb*, a monthly magazine that gave voice to the Young Tunisian's desires to reform the colonial system and advocated for self-determination by the end of the First World War.[74]

By constantly seeking the origins of Tunisian, Algerian, and Moroccan nationalism in France and the Maghreb, historians have almost forgotten the important role that the former imperial capital, Istanbul, played as a political center where activists crystallized their national and separatist agendas.

Fierce Competition Between Tunisian Notables

After the First World War, the Bu Hajib family underwent a third and final period of change with the impressive professional advancement of Salim's son Khalil (1863–1942). In his youth, Khalil became Husayn's intimate associate. After attending Sadiki College from 1875 to 1880, Khalil, together with his brother 'Amr, went to France, first in 1880 and then in 1882, to pursue their studies, and from there sometimes visited Husayn in Tuscany.[75] Their brother Husayn even bore the first name (*ism*) of the family's benefactor. The three of them, together with their younger sibling Ahmad, were also linked to Husayn through their management of the La Goulette *awqāf*.

Husayn Bu Hajib (1872–1946) had a brilliant career. After graduating from Bordeaux University, in 1904 he started working as a doctor at Sadiki Hospital (now known as the Aziza Othmana Hospital), located in the historical center of Tunis. He liked to present his work as a way of helping the dispossessed, especially those from Tripoli.[76] In 1905, he set up an association, *La Musulmane* (the Muslim League), which promoted physical exercise, something his father Salim Bu Hajib had valued.[77] However, his brother Khalil soon became even more successful. He had started his career in 1889, when he was nearly twenty-six, working as an interpreter in the "translation section," which connected the office of the Tunisian prime minister with the French colonial administration.[78] Until the end of the First World War, he mainly held second-rank positions in the Tunisian government. In 1912, the forty-nine-year-old Khalil was appointed head of the investigations section at the Ministry of Justice before being chosen to head the municipality of Tunis (*shaykh al-madīna*).[79] In 1926, after a long career in the Tunisian administration, the sixty-two-year-old became grand vizier (or prime minister) to the Tunisian *bey*, a prestigious position that he held through the early 1930s.[80]

The professional advancement of Husayn and Khalil was indeed related to their education in the French system, which was very unusual for Tunisian Muslims of the period. Salim Bu Hajib's sons belonged to a very privileged social and economic elite. In 1883, a total of three thousand students attended twenty-five "modern" European-Tunisian schools that did not follow the Islamic curriculum of the Quranic school. Two-thirds of these students were Europeans or Tunisian Jews. Only one thousand of them were Muslims, less than 0.1 percent out of a population of one to 1.5 million Muslim inhabitants in Tunisia at the time. Less than three decades later, in 1911, twelve thousand Muslim students attended these schools, although they still represented only one-third of the total student population and 0.6 percent of the total Muslim population.[81]

But the gradual divergence in the careers of the two brothers—and Khalil's impressive political achievement—stemmed from other causes as

FIGURE 6.2 Portraits of Princess Nazli (private collection of Rukiye Kuneralp and Staffordshire Record Office)

well. Khalil's rise to higher social positions than Husayn was also due to his status as a notable, acquired on marrying an Egyptian princess, Nazli (figure 6.2), niece of Khedive Isma'il, ruler of Egypt from 1863 to 1879.[82]

Khalil and Nazli married in 1897–98.[83] Khalil was thirty-four at the time, ten years younger than Nazli. Their union was another unexpected consequence of Husayn's exile in Florence. While married to the diplomat and former Ottoman minister Khalil Sharif Pacha (who died in 1879), Nazli had become acquainted with the circle of Tunisian exiles centered on Husayn and Bayram V during her travels in France and Italy.[84]

FIGURE 6.2 (*cont.*)

This cosmopolitan woman, who knew Arabic, Turkish, French, and English, apparently fell for Khalil while visiting her sister, Ruqaya Hanım, in Tunis. Ruqaya was the daughter-in-law of the unsavory character Mahmud Ibn ʻAyyād, encountered in chapter 3, the tax collector who had left Tunis for Paris and later Istanbul after having ruined the Tunisian state tax farms in his charge.[85]

On marrying Khalil, Princess Nazli brought to Tunis the salon life she had led at her residence in Cairo, where she hosted important figures such as Muslim scholar Muhammad ʻAbduh; writers Ibrahim al-Muwaylihi (1846–1906) and his son Muhammad (1868–1930); lawyer and soon-to-be political leader Saʻd Zaghlul (1859?–1927); lawyer, writer, and pioneer of

Egyptian feminism Qasim Amin (1863–1908).[86] In Cairo, her salon had been so influential as to attract criticism: Egyptian nationalist Muhammad Farid, for instance, denigrated Nazli, describing her "as receiving men in European fashion," arranging "musical evenings at her home," and drinking "lots of wine and especially champagne during the meal or indeed even before every meal."[87] Far from Cairo, the house of Princess Nazli and Khalil Bu Hajib became a crucible for intellectual debate, frequented by Tunisian scholars and famous travelers until Nazli passed away in 1913.

Even after his wife's death, Khalil kept on being promoted in the Tunisian administration as he appeared to be very loyal to the French colonial regime. Indeed, after 1912, he became public prosecutor at the French Court of First Instance in Tunis. However, his obedience to France was not the only factor that drove his advancement. Khalil had managed to climb the political ladder despite a lively competition for titles and honors among members of the Tunisian sovereign's entourage in Tunis. The ministers and grand viziers to the *bey* admittedly had very limited power over the true course of events in Tunisia; the French heads of ministerial services held their authority in check. But notables and their descendants constantly jockeyed for prestigious positions in the government hierarchy nevertheless. For instance, upon becoming the Tunisian grand vizier, Khalil Bu Hajib competed fiercely with a justice minister, Tahar Khayr al-Din Bacha. Tahar was the son of former prime minister Khayr al-Din, longtime patron of Husayn. He had been educated at the Galatasaray High School in Istanbul before becoming aide-de-camp to Sultan Abdülhamid II in 1893. Unable to find a stable position after the 1908 Young Turk revolution, Tahar made up his mind to go back to Tunis, where he quickly found an administrative post.[88]

The competition within the Tunisian *beys'* entourage under the French protectorate in the 1910s and 1920s has often been underestimated. These circles of provincial notables, their rivalries, and even their composition merit further analysis. They should not be seen only as a collaborationist elite from the old regime, swept away in the 1950s by the independence of Tunisia and the nationalist government led by president Habib Bourguiba. Indeed, each of these statesmen reacted to French domination in his own way, collaborating with or resisting the protectorate administration. Resistance and collaboration are certainly crucial terms for analyzing how this Tunisian elite, and the Bu Hajib household in particular, adapted to the colonial era. But their actions were intrinsically related to fierce social competition both among Muslim officials—as seen in Khalil Bu Hajib's career path—and among Muslim scholars, as shown in Shaykh Salim Bu Hajib's case.[89] Moreover, their interactions with French authorities did not prevent this Tunisian elite, or at least, the Bu Hajib household, from maintaining connections with Ottoman centers in Istanbul and Egypt until the First World War.

From this perspective, paying close attention to one of the final aftereffects of the Husayn affair—namely, the changes wrought in the Bu Hajib household from 1890 onward—reveals the complex web of connections spun by Tunisian elites, who not only collaborated with or resisted French colonialism, but who also needed to take into account various social forces such as their Tunisian competitors and their Middle Eastern contacts. Additionally, focusing on how the Bu Hajib family adapted to their circumstances shows that social and political structures such as the household, which have been so useful in enhancing our understanding of Ottoman societies, are still relevant to revising the history of post-Ottoman North African societies during the colonial period.[90]

Finally, and more broadly, by assessing the various outcomes of the conflicts over Husayn's estate from the perspective of the French state, and by following the specific history of a family, this last chapter proves that litigation in the context of legal pluralism did not merely shape the colonial order.[91] Litigants did not fuel further disputes just for the sake of keeping some room to maneuver.[92] Legal disputes also resulted in the persistence of local political, social, and legal cultures, such as the Ottoman-North African provincial practices that the Bu Hajib maintained during the colonial period.

Conclusion

Local and Imperial Histories of the Maghreb

THIS BOOK has used Husayn's life and the history of his estate after his death as a means to go beyond understanding the modern history of the Maghreb solely through the prism of European colonization. But what historical alternatives does this case offer? I contend that Husayn's story brings to light the local and imperial dimensions that complement the colonial version of this history. By local, I mean the multiple places—whether Florence or Livorno, Tunis or Istanbul—where various political and commercial interests met. By imperial, I mean not only the Ottoman Empire and its profound influence on the Maghreb, but the entirety of the European and Mediterranean spaces in which this story's actors competed and operated. By highlighting how the Ottoman Empire, the wider Muslim world, and individual households remained significant foci for the hopes and actions of Tunisians and other North Africans at the height of the colonial period, this study opens up new lines of inquiry and, hopefully, new historical interpretations.[1]

The Major Effects of Legal Actions, Claims, and Litigation

Among these lines of inquiry, this specific case has zoomed in on how Husayn and his entourage pressed claims, launched litigation, and waged disputes. The cases involving colonized subjects, which have attracted much interest in studies of legal pluralism, are fascinating for their various important social and political effects. Legal disputes in a colonial context did not, I argue, contribute only to the structuring of the colonial legal systems or, conversely, to tearing those same systems apart.[2] As we saw in chapter 3, the conflicts over Mahmud Ibn 'Ayyad's and Nessim Scemama's estates strengthened financial and legal relations between Tunis, France, and Italy both before and during the colonial domination of Tunis, from the 1850s to the 1910s. Though colonized subjects did not completely trust European legal institutions, I have also underlined the extent to which claimants would use litigation to recover or even gain agency in asymmetrical situations, such as when Husayn decided to provoke new legal problems after the French conquest in 1881, when he realized that he could no longer rely on the material and diplomatic resources of the Tunisian state.

Close study of the legal disputes surrounding Husayn's estate has also revealed various forms of clandestine action that, by definition, appear only fleetingly, if at all, in the archives of European colonial administrations. Tunisian sovereign 'Ali Bey, for instance, retained secret connections with Istanbul and with Léon Elmilik, despite the French colonial administration's warnings against such relations. These low-intensity conflicts, which are indeed less striking than open revolts and rebellions, point to the multifaceted meanings of colonized subjects' actions, which went beyond straightforward resistance or collaboration.

On both shores of the Mediterranean, these never-ending conflicts had yet another major consequence: they contributed to the proliferation of legal, religious, and administrative sources in Arabic, Judeo-Arabic, Ottoman Turkish, and European languages.[3] Studies on the colonization of North Africa often ignore, underestimate, or wrongly deny the profusion of writing in North African societies. Husayn's case not only confirms that the increased number of claims and disputes on either side of the Mediterranean generated a flood of documents, but it also reminds us that these documents combined European and Arabic languages and underlines the increasing pervasiveness of sources written in European languages on Islamic legal issues.

Muslim Households Versus the Colonial State

This study did not examine the actions of Husayn and his entourage solely through the lens of legal disputes and court proceedings. Husayn's case also highlights the lasting significance of households, families, and lineage in the modern and colonial history of Tunisia. Historians of the Ottoman Empire have already demonstrated the importance of households and lineage in their reassessment of politics and gender during the early modern period, but this rich scholarship has not yet significantly influenced our approach to the Maghreb during colonial times.[4] North African households are still viewed from a distance, as if over the shoulder of European administrators, as inert groups deprived of any historical weight.

However, as the case of the Bu Hajibs indicates, such households could retain some local influence. In a way, Husayn created a particular household that brought together the two girls and their mothers. Afterward, within the Bu Hajib household, Angiolina Bertucci enjoyed far greater opportunities than women of similar background, like Eva Keusch, who failed to gain any significant support from the French colonial authorities. Outside of Tunisia, the Bu Hajibs also succeeded in maintaining relationships with dignitaries from Istanbul and Cairo until the 1920s through two important women in the household: Farida (née Maria/Myriam) and Princess Nazli. Similarly, a far larger household, the palace of the Tunisian *beys*, was also able to continue informal policy-making. Despite being closely overseen by the *résidence générale*, ʻAli Bey could still rely on his sons and intimate acquaintances to promote his family interests. Throughout the colonial period, the palace was still a place of competition for notables, such as the Bu Hajibs, seeking positions and honors.

The Lost Territories of North African Historiography

Finally, the local and imperial dimensions of Husayn's story do not merely cast light on social institutions (such as households) and analytical tools (such as legal disputes) that are useful for reassessing the beginning of the colonial period in Tunisia and the lasting Ottoman influences on North Africa. They also suggest that we need to take into account venues outside of North Africa that were crucial for Maghrebis during the colonial period—territories that North African historiography has lost sight of in its focus on the relations between the North African colonies and the French mainland. True, North Africans invested less in Europe than Europeans did in the Maghreb. Still, despite this major difference, the importance of

Tuscany and Istanbul for social networks involving various North Africans across the Mediterranean should be stressed.

Tuscany and the heart of the Ottoman Empire were not merely havens, social hubs, or places of mediation. Outside of the Maghreb, Husayn and other exiled Tunisians reconceptualized their understanding of themselves as Muslims, Ottoman citizens, and colonized peoples. These places were important arenas for debating issues such as nationality and kinship. Istanbul was also a major venue where calls for North African independence were voiced as early as the first decade of the twentieth century. It would therefore be reductive to consider the issue of nationality in North Africa solely as a colonial invention, or to think (and teach) that calls for independence in the Maghreb were first voiced by North African workers and students in Europe during the interwar period. In both Western Europe and Istanbul, various versions of North African nationalism were born in exile, out of homesickness.

At the same time, this book does not cultivate a nostalgia for the Ottoman Empire. I do not argue that Istanbul or even Tuscany allowed complete freedom of expression to North African subjects, nor that they offered greater opportunities than the colonial states. During the colonial period, Ottomans and Italians alike sought to control Tunisian citizens. Istanbul sought to re-Ottomanize Tunisians. The Ottoman consul in Florence denied any form of Tunisian autonomy and any right for Tunisian (as opposed to Ottoman) diplomatic representation in Italy or elsewhere. For their part, Italian officials and Italian courts ignored the legal claims of Tunisian Jews and Muslims if these claims did not fit with the Italian political agenda in North Africa. Furthermore, other Mediterranean locales, not just Italy, deserve attention as crucial spaces for Maghrebis—especially Cairo, which increasingly became a center of intellectual and political exchange with Tunisian elites during the colonial period, and Sicily, where the vast majority of Italians settling in Tunisia during the colonial period originated.

Other Chronologies

This book does not deny the major impact of European colonialism, its violence, and its disastrous effects. Rather, I have sought to further our understanding of the colonial period by considering both what preceded and what lasted into this era. Such an approach brings other chronologies into view.[5] I do not consider the French conquest of Algeria in 1830 or the French invasion of Tunisia in 1881 to be the only milestones in the history of this region. I have argued that, in parallel to the French conquest of Algeria, the era of Ottoman political reforms during the second half of the nineteenth century needs to be reappraised as a watershed in its own right, affecting, as it did, the lives of Tunisian-Ottoman dignitaries, such as Husayn, and their entourages for decades. It was a time when the rights of former slaves converged with those of freeborn men and the rights of Christian and Jewish minorities started to align with those of Muslims (before Muslims' rights were restricted during colonial times).

This book has also shown to what extent the 1920s could be seen as a turning point, for it was during this decade that the Ottoman legacy and references to that Muslim empire started to run dry. It was also during the 1920s that the political descendants of Ottoman dignitaries such as Husayn, whom the colonizers had labeled Young Tunisians and Young Algerians, gave way to movements based on political parties. People no longer looked to Istanbul and its deposed sultans. They dreamed of Cairo, Mecca, and other places. They started to listen to national leaders. The time had come for other hopes.

Select Glossary

ʿadl (pl. *ʿudūl*): witness, notary
ʿālim, *ʿulamāʾ*: Islamic scholars, men of religion
ʿaqd: rental agreement
bayt al-māl: public treasury tasked with collecting revenue left unclaimed or without heir
bey, *beys*: governors of Ottoman Tunisia from the mid-seventeenth century who ruled in the name of the Ottoman sultan until 1881, and then under French colonial domination until 1957
beylik: lands under the control of local *beys*
Bilād al-Shām: the area encompassing Syria and Palestine
carabinieri: Italian armed forces
faqīh: man of law
fatwā: legal consultation
firmān: Ottoman imperial decree from the sultan
Garp ocakları: western provinces of the Ottoman Empire
ghurba: emigration, estrangement
Ḥanafī (*madhhab*): school of Islamic jurisprudence favored by the Ottoman elites in North Africa
ḥalqa: circle of students
hijra: emigration of the prophet Muhammad from Mecca to Medina in 622
ḥubūs: type of property set aside from the land market to support the heirs of the *ḥubūs* founder and, eventually, public and pious causes
ʿilm: knowledge
imām: man who leads Muslim prayers
inzāl: a lease in perpetuity on part of a *ḥubūs*
iṣlāḥ: reform
jamʿiyya al-ḥubūs: administration in charge of public *ḥubūs*

jazīrat al-Maghrib: island of the Maghreb, the lands between the Mediterranean Sea and the Sahara
kapı kethüdası: administrator who introduced Tunisian emissaries to Ottoman ministers
kātib: secretary, scribe
Maghreb: western part of the Muslim lands
majlīs al-shar'i: Islamic court
Mālikī (*madhhab*): school of Islamic jurisprudence predominant in North Africa
mamlūk: a slave converted to Islam and/or a servant raised within a household and promoted to an administrative or military position
mawlā: refers both to the slaveholder and to manumitted slaves
milk: private property
mu'addib: instructor, teacher
muftī: Islamic legist trained to issue legal opinions (*fatwās*)
nazīla, *nawāzil*: affair, case
niẓām al-jadīd: non-slave army established during the Ottoman reforms
pasha: honorary title granted to governors, generals, and other prominent dignitaries in the Ottoman Empire
qāḍī: judge in charge of applying *Shari'a* (Muslim) law
qā'id: town, district or village leader
résidence générale: institution representing France in Tunis
riḥla: travel and travelogue
risāla: missive, open letter, epistle
shahāda: oral testimony
shāshiya or *fez*, *ṭarbūsh*: red woolen caps sold across the Ottoman Empire and manufactured primarily in Tunis; part of the official uniforms during the Ottoman reform era
shaykh, *shuyūkh*: respected elder or leader
shaykh al-islām: highest Islamic legal authority in colonial Tunis
sulta: power
tanẓīmāt: Ottoman reforms, reorganizations
turbat al-bey: mausoleum of the Tunisian governors in the old city (*medina*) of Tunis
'ulamā': see *'ālim*
umma: the community of Muslims
wakīl, *ūkalā'*: person of trust, representative; traders in charge of Tunisian business in Istanbul
walā': contractual relationship between masters and former slaves
waqf/awqāf: see *ḥubūs*

Notes

Introduction

1. Husayn Ibn ʿAbdallah, Letter 256, August 6, 1882, in *Rasāʾil Ḥusayn ilā Khayr al-Dīn*, ed. Aḥmad ʿAbd al-Sālam (Carthage: Bayt al-Ḥikma, 1992), 3:122.

2. Husayn, Letter 280, May 11, 1886, in *Rasāʾil*, 3:175–77; and Tunisian National Archives [hereafter TNA], C. 11, d. 97, doc. 245, Husayn Ibn ʿAbdallah to Giaccomo Guttierez, Bellosguardo, Villa Bombici, June 4, 1886.

3. Husayn, Letter 290, April 3, 1887, in *Rasāʾil*, 3:195–96; TNA, C. 11, d. 109, doc. 8102, Salim Bu Hajib to the Tunisian Prime Minister, May 7, 1887.

4. Ministère des Affaires étrangères (MAE) (Nantes), 227 PO/1/100bis, Husayn's legacy, August 22, 1888.

5. Jean Ganiage, *Les Origines du Protectorat français en Tunisie (1861–1881)* (Paris: Presses universitaires de France, 1959); and Xavier Yacono, *Histoire de la colonisation française* (Paris: Presses universitaires de France, 1973).

6. Robert Mantran, *Inventaire des documents d'archives turcs du Dar El-Bey (Tunis)* (Paris: Presses universitaires de France, 1961). About the Berber sources, see Georges-Henri Bousquet, *Les Berbères*, 4th ed. (Paris: Presses universitaires de France, 1974), 87.

7. Joseph and Zivia Tobi, *Judeo-Arabic Literature in Tunisia, 1850–1950* (Detroit, MI: Wayne State University Press, 2014); Robert Attal, *Kitve-ʿet ṿe-ʿitonim Yehudiyim bi-Tsefon Afriḳah (Périodiques juifs d'Afrique du Nord)* (Tel Aviv: Tel Aviv University Press, 1996).

8. Mohamed Chérif Sahli, *Décoloniser l'histoire, introduction à l'histoire du Maghreb* (Paris: F. Maspero, 1965); and Abdallah Laroui, *The History of the Maghrib: An Interpretive Essay* (Princeton, NJ: Princeton University Press, 1977).

9. Edmund Burke III, "Towards a History of the Maghrib," *Middle Eastern Studies* 11, no. 3 (1975): 306–23.

10. Lucette Valensi, *Fellahs tunisiens: L'économie rurale et la vie des campagnes aux 18e et 19e siècles* (Paris: Mouton, 1977).

11. According to the National Library of France's website, from 2000 to 2018, some 3,520 books were published about Algeria (the bulk of them in French and dealing with the colonial history), 2,173 about Morocco, and 1,078 related to Tunisia (here again with quite an emphasis on the colonial period).

12. Emmanuelle Saada, *Les Enfants de la colonie: Les métis de l'empire français entre sujétion et citoyenneté* (Paris: Découverte, 2007); Patrick Weil, *Qu'est-ce qu'un Français? Histoire de la nationalité française depuis la Révolution* (Paris: Grasset, 2002); Laure Blévis, "La Citoyenneté française au miroir de la colonisation: Étude des demandes de naturalisation des 'sujets français' en Algérie coloniale," *Genèses* 53, no. 4 (2003): 25–47.

13. James McDougall, *History and the Culture of Nationalism in Algeria* (Cambridge: Cambridge University Press, 2006), 31.

14. Sylvie Thénault, "L'État colonial. Une question de domination," in *Les Empires coloniaux: XIXe–XXe siècle*, ed. Pierre Singaravelou (Paris: Points, 2013), 228.

15. Julia Ann Clancy-Smith, *Mediterraneans: North Africa and Europe in an Age of Migration, c. 1800–1900* (Berkeley: University of California Press, 2011), 340.

16. See also Ross E. Dunn, *Resistance in the Desert: Moroccan Responses to French Imperialism 1881–1912* (Madison: University of Wisconsin Press, 1977).

17. Most scholarship of the early modern period in the Maghrib still tends to be poorly disseminated in other countries, despite the importance of the insights it offers. See, for example, Abdelhamid Hénia, *Propriété et stratégies sociales à Tunis, XVIe–XIXe siècles* (Tunis: Université Tunis I, 1999); Sami Bargaoui, "'Quando dire è fare,' ovvero come ripetizione giuridica diventa una fonte storica," *Quaderni storici* 129, no. 3 (2008): 593–620; Fatma Ben Slimane, *Al-arḍ wa al-huwiyya, nushū' al-dawla al-turabiyya fī Tūnis 1881–1974* (Tunis: University of Tunis, 2009).

18. Valensi, *Fellahs tunisiens*; Abdelhamid Hénia, *Le Grid: Ses rapports avec le beylik de Tunis (1676–1840)* (Tunis: Publications de l'Université de Tunis, 1980); Jocelyne Dakhlia, *L'Oubli de la cité: La mémoire collective à l'épreuve du lignage dans le Jérid tunisien* (Paris: Découverte, 1990).

19. Edmund Burke, "Theorizing the Histories of Colonialism and Nationalism in the Arab Maghrib," *Arab Studies Quarterly* 20, no. 2 (1998): 5–19; and Daniela Melfa, "À la recherche des voix des 'subalternes': Le cas de Thibar vu à travers les archives des Missionnaires d'Afrique," in *Chantiers et défis de la recherche sur le Maghrib contemporain*, ed. Pierre-Robert Baduel (Paris: Karthala, 2009), 135–58.

20. Isabelle Grangaud, "Affrontarsi in archivo. Tra storia ottomana e storia coloniale (Algeri 1830)," *Quaderni storici* 129, no. 3 (2008): 621–52.

21. Edmund Burke, *Prelude to Protectorate in Morocco: Precolonial Protest and Resistance, 1860–1912* (Chicago: University of Chicago Press, 1976); and Allan Christelow, *Algerians Without Borders: The Making of a Global Frontier Society* (Gainesville: University Press of Florida, 2012).

22. Clancy-Smith, *Mediterraneans*; Mary Dewhurst Lewis, *Divided Rule: Sovereignty and Empire in French Tunisia, 1881–1938* (Berkeley: University of California Press, 2013).

23. Nabil Matar, *Islam in Britain, 1558–1685* (Cambridge: Cambridge University Press, 1998); Jocelyne Dakhlia and Bernard Vincent, eds., *Les Musulmans dans l'histoire de l'Europe* (Paris: Albin Michel, 2011).

24. Joshua Schreier, *The Merchants of Oran: A Jewish Port at the Dawn of Empire* (Stanford, CA: Stanford University Press, 2017); and Jessica M. Marglin, "The Two Lives of Mas'ud Amoyal: Pseudo-Algerians in Morocco, 1830–1912," *International Journal of Middle East Studies* 44, no. 4 (2012): 651–70.

25. Manuel Borutta et Sakis Gekas, "A Colonial Sea: The Mediterranean, 1798–1956," *European Review of History: Revue européenne d'histoire* 19, no. 1 (2012): 1–13.

26. André Miquel, *La Géographie humaine du monde musulman jusqu'au milieu du 11e siècle. Géographie et géographie humaine dans la littérature arabe des origines à 1050* (Paris: École des hautes études en sciences sociales, 1967), 81, 280, 306–8, 529; and Christophe Picard and Jacques Alexandropoulos, "'Orient' et 'Occident' dans l'historiographie du Maghreb de la période coloniale à l'époque actuelle," in *Décoloniser l'histoire? De l'histoire coloniale aux histoires nationales en Amérique latine et en Afrique, XIXe–XXe siècles*, ed. Sophie Dulucq and Colette Zytnicki (Saint-Denis: Société française d'histoire d'outre-mer, 2003), 85–100.

27. Fatma Ben Slimane, "Between Empire and Nation-State: The Problem of Borders in the Maghrib," in *Mediterranean Frontiers: Borders, Conflict and Memory in a Transnational Era*, ed. Dimitar Bechev and Kalypso Nicolaïdis (London: Tauris Academic Studies, 2010), 45.

28. McDougall, *History and the Culture of Nationalism*, 36.

29. Maurizio Isabella and Konstantina Zanou, eds., *Mediterranean Diasporas: Politics and Ideas in the Long 19th Century* (London: Bloomsbury Academic, 2016), 3, 15. "The east-west axis of exchange, traditionally the most significant, ceded to—while never being eclipsed by—the imperialism of the vertical north-south axis." Clancy-Smith, *Mediterraneans*, 98.

30. Ilham Khuri-Makdisi, *The Eastern Mediterranean and the Making of Global Radicalism, 1860–1914* (Berkeley: University of California Press, 2010), 9.

31. Romain Bertrand, "Par-delà le grand récit de la Nation: L'identité nationale au prisme de l'histoire globale," *Savoir/Agir* 2 (2007): 51–60.

32. Abdeljelil Temimi, *Recherches et documents d'histoire maghrébine: La Tunisie, l'Algérie et la Tripolitaine de 1816 à 1871* (Tunis: Université de Tunis, 1971); Leon Carl Brown, ed., *Imperial Legacy: The Ottoman Imprint on the Balkans and the Middle East* (New York: Columbia University Press, 1996); Asma Moalla, *The Regency of Tunis and the Ottoman Porte, 1777–1814: Army and Government of a North-African Ottoman Eyālet at the End of the Eighteenth Century* (London: Routledge, 2004); Abderrahmane El Moudden and 'Abd al-Raḥīm Bin-Ḥāddah, *Al-'Uthmāniyyūn fī al-Maghrib min khilāl al-arshīfāt al-maḥalliyya wa-al-mutawasiṭiyya* (Rabat: Al-Ṭab'ah 1, 2005); Amy Aisen Kallander, *Women, Gender, and the Palace Households in Ottoman Tunisia* (Austin: University of Texas Press, 2013).

33. Ussama Makdisi, "Ottoman Orientalism," *American Historical Review* 107, no. 3 (2002): 768–96; Selim Deringil, "'They Live in a State of Nomadism and Savagery': The Late Ottoman Empire and the Post-Colonial Debate," *Comparative Studies in Society and History* 45, no. 2 (2003): 311–42; Zeynep Çelik, *Empire, Architecture, and the City: French-Ottoman Encounters, 1830-1914* (Seattle: University of Washington Press, 2008), 16, 247.

34. Amy Mills, James A. Reilly, and Christine Philliou, "The Ottoman Empire from Present to Past: Memory and Ideology in Turkey and the Arab World," *Comparative*

Studies of South Asia, Africa and the Middle East 31, no. 1 (2011): 133–36; James McDougall, "Crisis and Recovery Narratives in Maghribi Histories of the Ottoman Period (ca. 1870–1970)," *Comparative Studies of South Asia, Africa and the Middle East* 31, no. 1 (2011): 137–48; Olivier Bouquet, "Old Elites in a New Republic: The Reconversion of Ottoman Bureaucratic Families in Turkey (1909–1939)," *Comparative Studies of South Asia, Africa and the Middle East* 31, no. 3 (2011): 588–600; Michael E. Meeker, *A Nation of Empire: The Ottoman Legacy of Turkish Modernity* (Berkeley: University of California Press, 2002).

35. M'hamed Oualdi, *Esclaves et maîtres: Les mamelouks des beys de Tunis du XVIIe siècle aux années 1880* (Paris: Publications de la Sorbonne, 2011).

36. M'hamed Oualdi, "Mamluks in Ottoman Tunisia: A Category Connecting State and Social Forces," *International Journal of Middle East Studies* 48, no. 3 (2016): 473–90.

37. Husayn's writings are still kept in the Tunisian National Archives (TNA) in Tunis, at the Archives of the French Ministry of Foreign Affairs (MAE) in Paris and Nantes, at the State Archives of Livorno, and in the Diplomatic Archives in Rome and London. Very little evidence has been found at the Başbakanlık Osmanlı Arşivi [the Ottoman Archives of the Prime Minister's Office] in Istanbul.

38. Jean-Claude Passeron and Jacques Revel, eds., *Penser par cas* (Paris: Éd. de l'École des hautes études en sciences sociales, 2005), 11.

39. Howard Saul Becker et Charles C. Ragin, eds., *What Is a Case? Exploring the Foundations of Social Inquiry* (Cambridge: Cambridge University Press, 1992), 217, 225.

40. Christian G. De Vito, "Verso una microstoria translocale," *Quaderni storici* 3 (2015): 815–33; John Paul A. Ghobrial, "The Secret Life of Elias of Babylon and the Uses of Global Microhistory," *Past & Present* 222 (2014): 51–93; Sanjay Subrahmanyam, "Beyond Incommensurability: Understanding Inter-Imperial Dynamics," *Department of Sociology, UCLA, Theory and Research in Comparative Social Analysis* 32 (2005): 1–38. For examples of these global biographies, see Linda Colley *The Ordeal of Elizabeth Marsh: A Woman in World History* (New York: Pantheon, 2007); Natalie Zemon Davis, *Trickster Travels: A Sixteenth-Century Muslim Between Worlds* (New York: Hill and Wang, 2006); Mercedes Garcia-Arenal, *A Man of Three Worlds: Samuel Pallache, a Moroccan Jew in Catholic and Protestant Europe* (Baltimore, MD: Johns Hopkins University Press, 2003); Emma Rothschild, *The Inner life of Empires an Eighteenth-Century History* (Princeton, NJ: Princeton University Press, 2011).

41. Frederick Cooper, "What Is the Concept of Globalization Good For? An African Historian's Perspective," *African Affairs* 100 (2001): 189–213.

42. Romain Bertrand, "Au milieu du gué. Situation(s) de l'anthropologie historique du fait colonial," *Genèses* 97 (2014): 146.

43. E. Natalie Rothman, *Brokering Empire: Trans-Imperial Subjects Between Venice and Istanbul* (Ithaca, NY: Cornell University Press, 2012), 7.

44. Francesca Trivellato, "Is There a Future for Italian Microhistory in the Age of Global History?" *California Italian Studies* 2 (2011): 16.

45. I only partially follow Clancy-Smith, who reconstitutes the trajectory of Husayn's friend Khayr al-Din in one chapter of her book ("Chapter 9: Khayr al-Din al-Tunisi and a Mediterranean Community of Thought," in *Mediterraneans*, 315–41).

46. André Nouschi, *Enquête sur le niveau de vie des populations rurales constantinoises de la conquête jusqu'en 1919: Essai d'histoire économique et sociale* (Paris: Presses universitaires de France, 1961); and Jean Poncet, *La Colonisation et l'agriculture européennes en Tunisie depuis 1881, étude de géographie historique et économique* (Paris: Imprimerie nationale, 1961).

47. Kenneth M. Cuno, "Was the Land of Ottoman Syria Miri or Milk? An Examination of Juridical Differences Within the Hanafi School," *Studia Islamica* 81 (1995): 121–52; Nicolas Michel, *L'Égypte des villages autour du seizième siècle* (Louvain: Peeters, 2018); Kemal H. Karpat, "The Land Regime, Social Structure, and Modernization in the Ottoman Empire," in *Studies on Ottoman Social and Political History: Selected Articles and Essays* (Leiden: Brill, 2002), 327–51; Huri Islamoğlu, "Modernities Compared: State Transformations and Constitutions of Property in the Qing and Ottoman Empires," *Journal of Early Modern History* 5, no. 4 (2001): 353–86; Ekaterina A. Pravilova, *A Public Empire: Property and the Quest for the Common Good in Imperial Russia* (Princeton, NJ: Princeton University Press, 2014).

48. Sami Bargaoui, Isabelle Grangaud, and Simona Cerutti, "Introduction," in *Appartenance locale et propriété au nord et au sud de la Méditerranée* (Aix-en-Provence: Institut de recherches et d'études sur le monde arabe et musulman, 2015), https://books.openedition.org/iremam/3415.

49. Luc Boltanski and al., eds., *Affaires, scandales et grandes causes: De Socrate à Pinochet* (Paris: Stock, 2007).

50. Mostafa Minawi, *The Ottoman Scramble for Africa: Empire and Diplomacy in the Sahara and the Hijaz* (Stanford, CA: Stanford University Press, 2016).

51. Bargaoui Cerutti, and Grangaud, "Introduction."

52. Daniel Schroeter, "Identity and Nation: Jewish Migrations and Intercommunity Relations in the Colonial Maghrib," in *La Bienvenue et l'adieu: Migrants juifs et musulmans au Maghrib, XVe–XXe siècles* (Casablanca: Karthala, 2012), 1:125–39.

53. Romain Bertrand, "Politiques du moment colonial. Historicités indigènes et rapports vernaculaires au politique en 'situation coloniale,'" *Questions de recherche* 26 (2008): 26.

54. Delphine Gardey, *Écrire, calculer, classer: Comment une révolution de papier a transformé les sociétés contemporaines, 1800–1940* (Paris: La Découverte, 2008); and Étienne Anheim and Olivier Poncet, "Fabrique des archives, fabrique de l'histoire," *Revue de Synthèse* 125, no. 1 (2004): 1–14.

55. De Vito, "Verso una microstoria translocale"; and Angelo Torre, "'Faire communauté': Confréries et localité dans une vallée du Piémont (XVIIe–XVIIIe siècle)," *Annales. Histoire, Sciences Sociales* 62, no. 1 (2007): 101–35.

56. Michel Camau, "Politique dans le passé, aujourd'hui au Maghrib," in *Les Trajectoires du politique. Volume II, la greffe de l'Etat*, ed. Jean-François Bayart (Paris: Karthala, 1994), 65.

57. Jocelyne Dakhlia, "Muslims in France: History Under the Carpet," in *Empire Lost: France and Its Other Worlds*, ed. Elisabeth Mudimbe-Boyi (Lanham, MD: Lexington Books, 2009), 25–35.

58. Jocelyne Dakhlia, *Tunisie, le pays sans bruit* (Arles: Actes sud, 2011).

1. Husayn

1. A small town located thirty miles north of Tunis.

2. Husayn Ibn 'Abdallah, Letter 281, June 7, 1886, in *Rasā'il Ḥusayn ilā Khayr al-Dīn*, ed. Aḥmad 'Abd al-Sālam (Carthage: Bayt al-Ḥikma, 1991–92), 3:178–79.

3. David Scott, *Conscripts of Modernity: The Tragedy of Colonial Enlightenment* (Durham, NC: Duke University Press, 2004), 7, 115; Talal Asad, "Conscripts of Western Civilization," in *Dialectical Anthropology: Essays in Honor of Stanley Diamond*, ed. Christine Ward Gailey (Tallahassee: University Press of Florida, 1992), 337.

4. Carter Vaughn Findley, *Ottoman Civil Officialdom: A Social History* (Princeton, NJ: Princeton University Press, 1989); Marc Aymes, *Un grand progrès sur le papier: Histoire provinciale des réformes ottomanes à Chypre au XIX*[e] *siècle* (Louvain: Peeters, 2010); M. Şükrü Hanioğlu, *A Brief History of the Late Ottoman Empire* (Princeton, NJ: Princeton University Press, 2008).

5. Sylvie Aprile, *Le Siècle des exilés: Bannis et proscrits de 1789 à la Commune* (Paris: CNRS, 2010), 285; and Daniel Rivet, *Le Maghreb à l'épreuve de la colonisation* (Paris: Hachette littératures, 2009), 153.

6. Ehud R. Toledano, *Slavery and Abolition in the Ottoman Middle East* (Seattle: University of Washington Press, 1998); Ehud R. Toledano, "Shemsigül: A Circassian Slave in Mid-Nineteenth-Century Cairo," in *Struggle and Survival in the Modern Middle East*, ed. Edmund Burke III (New York: I. B. Tauris, 1993), 59–74; Dawn Chatty, "Refugees, Exiles, and Other Forced Migrants in the Late Ottoman Empire," *Refugee Survey Quarterly* 32, no. 2 (2013): 35–52.

7. M'hamed Oualdi, *Esclaves et maîtres: Les mamelouks des beys de Tunis du XVII*[e] *siècle aux années 1880* (Paris: Publications de la Sorbonne, 2011), 97.

8. Ismael Musah Montana, *The Abolition of Slavery in Ottoman Tunisia* (Gainesville: University Press of Florida, 2013), 84, 97.

9. Toledano, *Slavery and Abolition*, 83, and especially 108.

10. Oualdi, *Esclaves et maîtres*, chap. 7.

11. Husayn, Letter 13, May 15, 1867, in *Rasā'il*, 1:50.

12. Husayn, Letter 13, in *Rasā'il*, 1:46, and Letter 89, July 28, 1873, in *Rasā'il*, 1:198. Despite his complaints about the Tunisian government's marginalizing him and otherwise treating him poorly, Husayn continued to define himself as a mamluk.

13. Husayn Pasha, "Selection XII: Husayn Pasha. The Abolition of Slavery," in *Modern Arab Thought: Channels of the French Revolution to the Arab East*, ed. Ra'īf Khūrī (Princeton, NJ: Kingston Press, 1983), 152.

14. National Archives and Records Administration (NARA), Record Group 59, Microfilm T303, US Consul in Tunis Amos Perry and the Secretary of State, Seward, August 24, 1865.

15. I want to thank Samuel Niu for making this suggestion and pointing to Thomas Jefferson, *Notes on the State of Virginia* (Philadelphia: Prichard and Hall, 1788), in *Documenting the American South*, accessed February 10, 2019, https://docsouth.unc.edu/southlit/jefferson/jefferson.html.

16. Husayn, "Selection XII: Husayn Pasha," 155.

17. Dror Ze'evi, "Slavery," in *The Oxford Encyclopedia of the Islamic World* (New York: Oxford University Press, 1995), 4:79–81.

18. Aḥmad Ṭawīlī, *Al-Jinirāl Husayn: ḥayāthuhu wa-athāruhu* (Tunis: Baladīyat Tūnis, 1994).

19. Mahmud 'Abd al-Mula, *Al-Madrasa al-Ḥarbiya bi-Bārdū* (Tunis: MTM, 2003); Ali Chenoufi, "Un Rapport inédit en langue arabe sur l'école de guerre du Bardo," *Cahiers de Tunisie* 95, no. 6 (1976): 45–118; M'hamed Oualdi, "A l'école des palais: Les maîtrises de l'écrit parmi les mamelouks des beys de Tunis, des années 1770 aux années 1860," *European Journal of Turkish Studies* 6 (2007), https://ejts.revues.org/1403.

20. Chenoufi, "Un Rapport inédit," 62–63, 108–9.

21. Anne-Laure Dupont, "An Expression of Pan-Islamism in Tunisia at the Beginning of the French Protectorate: The Critique of the Tijāniyya in the Riḥla al-ḥijāziyya by Muḥammad b. 'Uthmān al-Sanūsī (d. 1900)," in *Sufism, Literary Production, and Printing in the Nineteenth Century*, ed. Rachida Chih, Catherine Mayeur-Jaouen, and Rüdiger Seesemann (Würzburg: Ergon Verlag, 2015), 401–36; and 'Umar Bin Sālim, *Qābādū 1815–1871: ḥayātuhu wa-āthāruhu wa-tafkīruhu al-islāḥī* (Tunis: Tunis University, 1975), 24–47, 51–53.

22. Anonymous, "Le Bey de Tunis, l'Algérie, le Constitutionnel, la Vérité," *Bulletin de la société orientale* 4, no. 13–16 (1844): 91.

23. Jean Ganiage, *Les Origines du Protectorat français en Tunisie, 1861–1881* (Paris: Presses universitaires de France, 1959).

24. Mongi Smida, *Khereddine: Ministre réformateur, 1873–1877* (Tunis: Maison tunisienne de l'édition, 1971); Leon Carl Brown, *The Tunisia of Ahmad Bey, 1837–1855* (Princeton, NJ: Princeton University Press, 1974); Chenoufi, "Un Rapport inédit."

25. M'hamed Oualdi, "Provincializing and Forgetting Ottoman Administrative Legacies: Sons and Grandsons of Beys' Mamluks Facing French Administrators of Tunisia (1890s–1930s)," *Comparative Studies of South Asia, Africa and the Middle East* 34, no. 2 (2014): 418–31.

26. Husayn Ibn 'Abdallah, Letter 2, January 5, 1857, in *Rasā'il*, 1:30–31.

27. 'Abd al-Karīm al-'Alāqī, "Khayr al-Dīn wa al-Yahūd fī Tūnis fī al-Qarn al-Tāsi' 'ashar," in *Khayr al-Dīn al-Tūnisī* (Tunis: Maṭba'a al-Maghrib li-l-Nashr, 2011), 149–81; and Léon Bercher, "En marge du Pacte 'fondamental.' Un document inédit," *Revue tunisienne* 37 (1939): 67–86.

28. Muḥammad al-Hādī al-Maṭwī, "Al-Jinirāl Husayn yaktubu awwal qiṣṣa aṭfāl fī Tūnis," *al-Masār* 1 (1988): 114–20.

29. Tunisian National Archives (TNA), C. 11, d. 113, doc. 8791, French Consul De Laigue to French Foreign Affairs Minister Flourens, Florence, July 20, 1887.

30. Husayn Ibn 'Abdallah, Letter 3, July 15, 1859, in *Rasā'il*, 1:32–33; and Letter 65, June 19, 1872, 1:149.

31. Muḥammad Maḥfūdh, *Tarājim al-mu'allifin al-Tūnisiyyin* (Beirut: Dār al-Gharb al-Islāmī, 1982), 65–69; Husayn Ibn 'Abdallah, Letter 190, September 6, 1878, in *Rasā'il*, 2:178; and Letter 283, July 9, 1886, in *Rasā'il*, 3:183–84.

32. Ministère des Affaires étrangères (MAE) (Nantes), 227 PO/1/100bis, Husayn's legacy, October–November 1887: "un cahier manuscrit contenant des vers, notes littéraires."

33. Fatḥī Qāsimī *Al-Shaykh Muḥammad Bayram al-Khāmis: Hayātuhu wa-Fikruhu al-Iṣlāḥī* (Carthage: Bayt al-Ḥikma, 1990).

34. Umar Bin Sālim, *Qābādū 1815–1871: ḥayātuhu wa-āthāruhu wa-tafkīruhu al-islāḥī* (Tunis, Université de Tunis, 1975), 51–55. Salim Bu Hajib was introduced to the circle of Tunis scholars thanks to the highest religious dignitary in Tunis, the Hanafi Shaykh *al-islām* Muhammad Bayram IV, who placed his vast library at Salim's disposal (ʿAbd al-Qādir Vaḥḥah, *Al-Ḥaraka al-iṣlāḥīya al-zaytūnīya wa-al-madrasīya al-Tūnisīya: al-Shaykh Sālim Bū Ḥājib wa-manhajuhu al-iṣlāḥī, 1244–1342/1827–1924* [Tunis: al-Ṭabʿa 1, 2007], 44).

35. Anne-Laure Dupont, "De la Demeure du Califat aux 'découvertes parisiennes': Muhammad al-SanÛsî (1851–1900), un lettré réformiste tunisien à l'épreuve du protectorat français," in *Penser, agir et vivre dans l'Empire ottoman et en Turquie. Etudes réunies pour François Georgeon*, ed. Nathalie Clayer et Erdal Kaynar (Louvain: Peeters, 2013), 48; Ali Chenoufi, *Un Savant tunisien du XIXe siècle, Muḥammad as-Sanūsī* (Tunis: Tunis University, 1977), 25; Muḥammad al-Fāḍil Bin ʿĀshūr, *Arkān al-nahḍah al-adabīya bi-Tūnis* (Tunis, Maktabat al-Najāḥ, 1961), 29.

36. Husayn, *Rasāʾil*, 1:16; Khayr al-Dīn al-Tūnisī, *Essai sur les réformes nécessaires aux états musulmans* (Aix-en-Provence: Édisud, 1987); Arnold H. Green, *The Tunisian Ulama 1873–1915: Social Structure and Response to Ideological Currents* (Leiden: Brill, 1978), 90, 110; Majmaʿ al-Tūnisī lil-ʿUlūm wa-al-Ādāb wa-al-Funūn, *Al-Shaykh al-Muṣliḥ Sālim Bū Ḥājib wa-ishkālīyat al-ʿaṣr* (Carthage: Bayt al-Ḥikma, 2007), 17, 18.

37. Husayn, Letter 13, May 15, 1867, in *Rasāʾil*, 1:45.

38. Ṭawīlī, *Al-Jinirāl Husayn*, 19, 24; Ganiage, *Les Origines du Protectorat français*, 83.

39. Christiane Souriau-Hoebrechts, *La Presse maghrébine: Libye, Tunisie, Maroc, Algérie; évolution historique, situation en 1965, organisation et problèmes actuels* (Paris: Éditions du CNRS, 1969), 35. See also Ṭawīlī, *Al-Jinirāl Husayn*, 22.

40. Husayn Ibn ʿAbdallah, Report on the Tunisian bey's visit to Algiers, September 1860, in *Rasāʾil*, 1:225–26.

41. Bice Slama, *L'Insurrection de 1864 en Tunisie* (Tunis: Maison tunisienne de l'édition, 1967); and Kalifa Chater, *Insurrection et répression dans la Tunisie du XIXe siècle: La méhalla de Zarrouk au Sahel, 1864* (Tunis: Publications de l'Université de Tunis, 1978).

42. Husayn, Letter 13, May 15, 1867, in *Rasāʾil*, 1:45–51.

43. Ṭawīlī, *Al-Jinirāl Husayn*, 24; and Ganiage, *Les Origines du Protectorat français*, 83.

44. For the stay in New York and Washington, see Ibn ʿAbdallah Husayn, Letter 196, October 26, 1878, in *Rasāʾil*, 2:188.

45. NARA, Record Group 59, Microfilm T303, US Consul in Tunis Amos Perry to Secretary of State William H. Seward, September 20, 1867: "PS: I learn from a private source That General Heussein, a copy of whose general letter of introduction I forwarded to the department last spring, sailed from Liverpool in August. His plan was to land in Canada and proceed to the States. He is, with one exception, the most enlightened musulman person in Tunis."

46. NARA, Record Group 59, Microfilm T303, Perry to Seward, April 6, 1867.

47. NARA, New York, Passenger and Crew Lists (including Castle Garden and Ellis Island), 1820–1957, Microfilm Serial M237, 1820–97, list number 282, August 20, 1867.

48. Husayn Ibn ʿAbdallah, Letter 28, September 30, 1868, in *Rasāʾil*, 2:86.

49. MAE (Nantes), 227 PO/1/100bis, October–November 1887, Husayn's legacy.

50. Husayn Ibn ʿAbdallah, Letter 288, December 8, 1886, in *Rasāʾil*, 3:192.

51. Husayn Ibn ʿAbdallah, Letter 1, October 21, 1856, in *Rasāʾil*, 1:27–29; Letter 17, February 27, 1868, in *Rasāʾil*, 1:59–60; Letter 21, July 28, 1868, in *Rasāʾil*, 1:79–80; and Letter 34, February 22, 1871, in *Rasāʾil*, 1:102–5.

52. Husayn Ibn ʿAbdallah, Letter 79, August 31, 1872, in *Rasāʾil*, 1:173.

53. Husayn Ibn ʿAbdallah, Letter 164, September 4, 1877, in *Rasāʾil*, 2:134.

54. *Riḥla Tūnisī bi-Urūba bi-qalami al-farīq jinirāl Husayn wazīr al-ʿulūm bi-Tūnis.*

55. Husayn Ibn ʿAbdallah, Letter 290, April 3, 1887, in *Rasāʾil*, 3:197.

56. M. W. Daly and Carl F. Petry, *The Cambridge History of Egypt* (Cambridge: Cambridge University Press, 1998), 170, 395–96, 404.

57. Dupont, "De la Demeure du Califat," 52–54.

58. MAE (Courneuve), Bey and family, file 1, "Direction politique des protectorats" to the French Foreign Office, November 26, 1893.

59. Husayn Ibn ʿAbdallah, Letter 10, September 1868, in *Rasāʾil*, 1:41.

60. Nile Green, "Spacetime and the Muslim Journey West: Industrial Communications in the Making of the 'Muslim World,'" *American Historical Review* 118, no. 2 (April 2013): 401.

61. Husayn, Letter 1, in *Rasāʾil*, 1:20.

62. Husayn Ibn ʿAbdallah, Letter 169, November 28, 1877, in *Rasāʾil*, 2:149.

63. Husayn, Letter 190, in *Rasāʾil*, 2:178.

64. Husayn Ibn ʿAbdallah, Letter 87, October 16, 1873, in *Rasāʾil*, 1:214.

65. Catherine Brice, "Confiscations et séquestres des biens des exilés politiques dans les états italiens au XIXe siècle. Questions sur une pratique et projets de recherches," *Diasporas: Circulations, migrations, histoire* 23–24 (2014): 147–63.

66. TNA, C. 11, d. 97, General Husayn Ibn ʿAbdallah to Tunisian Prime Minister, December 12, 1881.

67. Beatriz Alonso Acero, *Sultanes de Berbería en tierras de la cristiandad: exilio musulmán, conversión y asimilación en la monarquía hispánica, siglos XVI–XVII* (Barcelona: Edicions Bellaterra, 2006).

68. Charles Robert Ageron, "Les Migrations des musulmans algériens et l'exode de Tlemcen (1830–1911)," *Annales* 22, no. 5 (1967): 1047–66; and Allan Christelow, *Algerians Without Borders: The Making of a Global Frontier Society* (Gainesville: University Press of Florida, 2012), 52.

69. Husayn Ibn ʿAbdallah, Letter 191, September 26, 1878, in *Rasāʾil*, 2:180; Andreas Tunger-Zanetti, *La Communication entre Tunis et Istanbul, 1860–1913: Province et métropole* (Paris: L'Harmattan, 1996), 56. Rustam passed away in Alexandria in November 1886; he had married Mahbuba, Khayr al-Dīn's daughter, in 1883.

70. Bin ʿĀshūr, *Arkān al-nahḍa*, 18; and Vaḥḥah, *Al-Ḥaraka al-iṣlāḥīya*, 40.

71. TNA, C. 11, d. 113, doc. 8941, French consul to résident général Massicault, Florence, December 30, 1887; and Tunger-Zanetti, *La Communication entre Tunis et Istanbul*, 56.

72. Dupont, "De la Demeure du Califat," 51.

73. TNA, C. 11, d. 97, doc. 19, Aviso da inserire nel *Giornale officiale* di Roma, 1881: "S. A. Il Bey di Tunisi ha revocato ogni e qualunque mandato che aveva in precedenza conferito a S.E. il generale di divisione Heussein per tutelare gli interessi del Governo Tunisino verso la successione del fu conte caïd Nissim Scemama defunto in Livorno il 24 Gennaio 1873."

74. TNA, C. 11, d. 97, doc. 178, Husayn Ibn ʿAbdallah to Giaccomo Guttierez, Florence, May 31, 1885.

75. TNA, C. 11, d. 97, doc. 209, Husayn Ibn ʿAbdallah to Giaccomo Guttierez, Nice, August 4, 1885.

76. MAE (Courneuve), CPC, file 630, Husayn Ibn ʿAbdallah to Khayr al-Din, October 19, 1879.

77. Green, "Spacetime and the Muslim Journey West," 401–29.

78. MAE (Nantes), 227 PO/1/100bis, October–November 1887, Husayn's legacy; Benoit Marpeau, *Gustave Le Bon: Parcours d'un intellectuel 1841–1931* (Paris: CNRS Editions, 2000), 77–79; Anne-Laure Dupont, *Ǧurǧī Zaydān (1861–1914): Écrivain réformiste et témoin de la renaissance arabe* (Damas: IFPO, 2006), 28–30, 427.

79. BL (Biblioteca Labronica), fondo Michel Ersilio, *Lettera del generale Heussein agli onorevoli avvocati componenti il collegio della difesa del governo di Tunis, (traduzione dall'arabo)* (Florence: Ricci, 1880), 26: "Voi ci chiamate dinanzi ai vostri tribunali, ma ci fate sapere che i vostri tribunali non debbono studiare le nostre leggi, né prende cognizione dei nostri costumi, né rispettare i nostri trattati, perché noi siamo piccoli, siamo deboli, siamo barbari, e siamo ridicoli."

80. Rashād al-Imām, "Tārīkh al-Wazīr Muṣṭafā b. Ismāʿīl," *al-Abḥāth* 17 (1969): 71–135.

81. Ṭawīlī, *Al-Jinirāl Husayn*, 28, 29; and MAE (Courneuve), CPC, file 630, Husayn's legacy, March–April 1884, Envelope V, Jamal al-Dīn to Husayn (three letters).

82. Ṭawīlī, *Al-Jinirāl Husayn*, 28, 29; MAE (Courneuve), CPC, file 630, Husayn's legacy, March–April 1884, Envelope V, Jamal al-Dīn to Husayn (three letters); MAE (Nantes), 227 PO/1/100bis, October–November 1887, Husayn's legacy, f. 103 v., "Photographie d'engagement en date du 23 chaoual 1300, par lequel Hussein subventionne un journal arabe que fera publier Brahim Elmosselhi."

83. Green, *The Tunisian Ulama*, 147.

84. Al-Afghānī, Jamāl al-Dīn and ʿAbduh, Muḥammad, *Al-ʿUrwa al-Wuthqā*, ed. Muḥammad Jamāl (Cairo: al-Maktaba al-ahliyya, 1927), 526–28.

85. Husayn Ibn ʿAbdallah, Letter 285, August 9, 1886, in *Rasāʾil*, 3:189.

86. Husayn, Letter 290, in *Rasāʾil*, April 3, 1887, 3:197.

87. Husayn, Letter 290, in *Rasāʾil*, 3:197.

88. Bercher, "En marge du Pacte 'fondamental,'" 75, 85.

89. Husayn Ibn ʿAbdallah, Letter 265, December 6, 1883, in *Rasāʾil*, 3:140; Letter 268, August 22, 1884, in *Rasāʾil*, 3:145; MAE (Nantes), 227 PO/1/100bis, October–November 1887, Husayn's legacy; TNA, C. 11, d. 105, doc. 7788, Léon Elmilik, credits and rights' assignments, Livorno, December 14, 1884.

90. TNA, C. 11, d. 113b, doc. 9153, David Santillana to Secretary General Régnault, Florence, May 4, 1888.

91. Habib Kazdaghli, "The Case of Tunisia," and Benjamin Stora, "The Crémieux Decree," in *A History of Jewish-Muslim Relations: From the Origins to the Present Day*, ed. Abdelwahab Meddeb and Benjamin Stora, trans. Jane Marie Todd and Michael B. Smith (Princeton, NJ: Princeton University Press, 2013), 316–19, 286–91.

92. TNA, C. 11, d. 100, doc. 7521, Eva Keusch to Résident général, Tunis, October 17, 1891.

93. Dorian Bell, *Globalizing Race: Antisemitism and Empire in French and European Culture* (Evanston, IL: Northwestern University Press, 2018), 91.

94. Bell, *Globalizing Race*, 15, 41, 91.

95. Archives of Livorno, Birth Certificate Register, certificate 43 (1882) and certificate 54 (1876).

96. Archives of Livorno, Birth Certificate Register, certificate 43 (1882) and certificate 54 (1876). See also MAE (Nantes), 227 PO/1/100bis, October–November 1887, Statement before a notary, November 13, 1880. Emma/Amina was recorded on the Livorno civil register under her father's name, al-Mu'addib, on April 15, 1882, according to a deed signed by Eva Keusch (TNA, C. 11, d. 100, doc. 7627, Florence, September 13, 1888).

97. TNA, C. 11, d. 100, doc. 7624, Agreement between David Santillana and, Bertrand Meuville, Florence, August 12, 1888.

98. TNA, C. 11, d. 100, doc. 7631, Richard Nicolas, December 1, 1888.

99. MAE (Nantes), 227 PO/1/100bis, October–November 1887, Husayn's legacy.

100. TNA, C. 11, d. 100, doc. 7684, Death Certificate, February 27, 1887.

101. TNA, C. 11, d. 100, doc. 7624.

102. TNA, C. 11, d. 100, doc. 7541, Richard Nicolas to résident general, Tunis, November 26, 1888.

103. TNA, C. 11, d. 100, doc. 7672, Eva Keusch to the Representative of the French Republic, Tunis, July 28, 1888.

104. MAE (Nantes), 227 PO/1/100bis, October–November 1887, Husayn's legacy.

105. TNA, C. 11, d. 113b, doc. 9116, Bertrand Meuville to Bey of Tunis, New York, February 27, 1888.

106. MAE (Nantes), 227 PO/1/100bis, October–November 1887, Husayn's legacy.

107. Lucilla Briganti, "L'emigrazione 'stagionale' dalla Toscana in Tunisia tra Ottocento e Novecento," in *Tunisia e Toscana*, ed. Vittorio Antonio Salvadorini (Pisa: Edistudio, 2002), 151.

108. MAE (Nantes), 227 PO/1/100bis, October–November 1887, Husayn's legacy.

109. MAE (Nantes), 227 PO/1/100bis, October–November 1887, Note "casa Giacosa, A. Bono" and "C. Baudino."

110. MAE (Nantes), 227 PO/1/100bis, October–November 1887, Husayn's legacy.

111. MAE (Nantes), 227 PO/1/100bis, October–November 1887, Diary 1884.

112. MAE (Nantes), Tunisian Protectorate, file: Bayram, 1TU/1V/1495, "Direction Politique" to "Président du Conseil," Tunis, February 23, 1885.

113. MAE (Courneuve), CPC, file 630, Turkhan Bey, Ottoman Ambassador in Berlin to Husayn, February 5, 1876.

114. Husayn Ibn 'Abdallah, Letter 289, December 30, 1886, in *Rasāʾil*, 3:193–94.

115. Mustafa's first wife passed away in 1888, and he repudiated his second wife. Note written about Mustafa b. Ismail by R. Catroux, "contrôleur civil," in MAE (Courneuve), CPC, file 627, Tunis, June 30, 1890.

116. Note on Husayn's real estate in TNA, C. 11, d. 113b, doc. 8957, La Goulette-Halq al-Wadi, f. 2^{r}: "mais ces deux filles (et le général montrait les deux mineures Amina et Meriem) nées chez moi et élevées par moi dans la religion musulmane, je ne puis les abandonner."

117. TNA, C. 11, d. 100, doc. 7624; and TNA, C. 11, d. 100, doc. 7524, Direction politique au Protectorat, September 26, 1887, Tunis. See also David Stephan Powers,

The Development of Islamic Law and Society in the Maghrib: Qāḍīs, Muftīs, and Family Law (Burlington, VT: Ashgate Variorum, 2011).

118. David S. Powers, "Adoption," in *Encyclopaedia of Islam, THREE*, ed. Kate Fleet, Gudrun Krämer, Denis Matringe, John Nawas, and Everett Rowson, http://dx.doi.org/10.1163/1573-3912_ei3_SIM_0304.

119. Husayn Ibn 'Abdallah, Letters 286–87, 7 dhū al-ḥijja 1303, September 6, 1886, in *Rasāʾil*, 3:190. Muslim scholars Salim Bu Hajib and Mustafa Ibn Ali Ridhwan are listed as witnesses of this repudiation.

120. Husayn Ibn 'Abdallah, Letters 286–87, in *Rasāʾil*, 3:190.

121. Husayn, Letter 285, in *Rasāʾil*, August 9, 1886, 3:186.

122. Rifaat Ali Abou El-Haj, "The Ottoman Vezir and Paşa Households 1683–1703: A Preliminary Report," *Journal of the American Oriental Society* 94, no. 4 (1974): 438–47; Jane Hathaway, *The Politics of Households in Ottoman Egypt: The Rise of the Qazdağlis* (New York: Cambridge University Press, 1997); Amy Aisen Kallander, *Women, Gender, and the Palace Households in Ottoman Tunisia* (Austin: University of Texas Press, 2013).

123. Piero Melograni and Lucetta Scaraffia, eds., *La Famiglia italiana dall'Ottocento a oggi* (Rome: Laterza, 1988); David I. Kertzer and Richard P. Saller, *The Family in Italy: From Antiquity to the Present* (New Haven, CT: Yale University Press, 1991); Beshara Doumani, "Endowing Family: Waqf, Property Devolution, and Gender in Greater Syria, 1800 to 1860," *Comparative Studies in Society and History* 40, no. 1 (1998): 3–41; Catherine Mayeur-Jaouen, "Les Familles au Moyen-Orient moderne et contemporain (XVIIe–XXIe siècle)," in *La Famille en islam: D'après les sources arabes* (Paris: Les Indes savantes, 2013), 391–489.

124. Ian Coller, *Arab France: Islam and the Making of Modern Europe, 1798–1831* (Berkeley: University of California Press, 2011); Noureddine Amara, "Les Nationalités d'Amîna Hanım. Une pétition d'hérédité à la France (1830–1896)," *Revue des mondes musulmans et de la Méditerranée* 137 (2015): 49–72.

2. Husayn's Wealth

1. Ministère des Affaires étrangères (MAE) (Nantes), 227 PO/1/100bis, 1887–89, Husayn's legacy, November 17–18, 1887, *Giornale del Popolo*; and Tunisian National Archives (TNA), C. 11, d. 113, doc. 8915, French Consul de Laigue to David Santillana, November 29, 1887, Florence.

2. Properties located in Ibn 'Abdallah and Jouggar.

3. M'hamed Oualdi, "L'âge du fez: du couvre-chef impérial au chapeau folklorique," in *Histoire du monde au XIXe siècle*, ed. P. Singaravélou and S. Venayre (Paris: Fayard, 2017), 476–80.

4. Sami Bargaoui, Isabelle Grangaud, and Simona Cerutti, "Introduction," in *Appartenance locale et propriété au nord et au sud de la Méditerranée* (Aix-en-Provence: Institut de recherces et d'études sur le monde arabe et musulman, 2015), https://books.openedition.org/iremam/3415.

5. Ekaterina A. Pravilova, *A Public Empire: Property and the Quest for the Common Good in Imperial Russia* (Princeton, NJ: Princeton University Press, 2014), 133.

Pravilova shows that in the context of imperial Russia, the legal construct of the idea of "public domain" and the promotion of private property in the second half of the nineteenth century affected material property, land, natural resources, water, and forestry. It also led to the redefinition of other, less material domains—notably, what ought to constitute the state, what was thought of as public property, and even ideas relating to artistic property and copyright.

6. M'hamed Oualdi, *Esclaves et maîtres: Les mamelouks des beys de Tunis du XVIIe siècle aux années 1880* (Paris: Publications de la Sorbonne, 2011), 247.

7. Charlotte Jirousek, "The Transition to Mass Fashion System Dress in the Later Ottoman Empire," in *Consumption Studies and the History of the Ottoman Empire, 1550–1922: An Introduction*, ed. Donald Quataert (Albany: State University of New York Press, 2000), 233.

8. Jirousek, "The Transition to Mass Fashion," 237.

9. Jirousek, "The Transition to Mass Fashion," 256. See also Leïla Blili, "Froufrous et bruissements: Costumes, tissus et couleurs dans la cour beylicale de Tunis au XIXe siècle," in *Trames de langue: Usages et métissages linguistiques dans l'histoire du Maghreb*, ed. Jocelyne Dakhlia (Paris: Maisonneuve et Larose, 2004), 223–39.

10. Florence National Archives (FNA), fasc. 2, rep. 632, October 17, 1887, "Verbale di rimozione e apposizione sigilli via Fra Bartolomeo 40."

11. MAE (Nantes), 227 PO/1/100bis, 1887–89, Husayn's legacy, May 19, 1988, List of items brought by the chancellor of the French consulate in Florence to the head of the auction house, Negri, Florence.

12. Krzysztof Pomian, *Collectionneurs, amateurs et curieux: Paris, Venise, XVIe–XVIIIe siècle* (Paris: Gallimard, 1987), 19, 35, 43; and Renata Ago, *Gusto for Things: A History of Objects in Seventeenth-Century Rome* (Chicago: University of Chicago Press, 2013), 7, 15, 30.

13. FNA, fasc. 2, rep. 632, October 17, 1887, "Verbale di rimozione e apposizione sigilli via Fra Bartolomeo 40."

14. MAE (Nantes), 227 PO/1/100bis, October–November 1887, Husayn's legacy, envelope containing four registers where a mention in Arabic on f. 69 states: "accounts of Miss Clarke, cooking expenses and other minor expenses from 1882 to the end of February 1885 including the rent of the house in Florence."

15. Arjun Appadurai, *The Social Life of Things: Commodities in Cultural Perspective* (Cambridge: Cambridge University Press, 1988), 30, 37.

16. Husayn Ibn ʿAbdallah, Letter 115, January 23, 1875, in *Rasāʾil Husayn ilā Khayr al-Dīn*, ed. Aḥmad ʿAbd al-Sālam (Carthage: Bayt al-Ḥikma, 1991–92), 2:50.

17. MAE (Nantes), 227 PO/1/100bis, October–November 1887, Husayn's legacy.

18. Pomian, *Collectionneurs, amateurs et curieux*, 49.

19. Michael Gilsenan, "Translating Colonial Fortunes: Dilemmas of Inheritance in Muslim and English Laws Across a Nineteenth-Century Diaspora," *Comparative Studies of South Asia, Africa and the Middle East* 31, no. 2 (2011): 356–57; MAE (Nantes), 227 PO/1/100, 1887–1888, Subfolder Heussein, June 29, 1887; Villa Sicard, Florence folders, subfolder Heussein June 29, [18]87, villa Sicard, in Giramonte, near the tower of Galla, Bellosguardo.

20. FNA, fasc. 5, rep. 635, October 17, 1887, "Verbale di remozioni sigilli sostituzione e consegna villa Sicard." See also TNA, C. 11, d. 113, doc. 8848, written

posthumously, before the chancellor of French consulate: "Trois malles, une caisse en bois blanc, deux valises en cuir jaune et cinq sacs de voyage en cuir noir de différentes grandeurs contenant des papiers faisant partie de la succession."

21. TNA, C. 11, d. 113, doc. 8811, French State councilor to résident général Massicault, August 27, 1887: "Je constate en terminant que . . ., main basse a été faite sur plus d'un objet de prix: ainsi des nombreuses décorations du défunt dont plusieurs pierres précieuses il n'a été retrouvé que les écrins."

22. TNA, C. 11, d. 113, doc. 8811.

23. MAE (Nantes), 227 PO/1/100bis, 1887–89, Husayn's legacy, *Giornale del Popolo*, November 17–18, 1887.

24. Fatiha Loualich, "'Alim Families in Seventeenth-Century and Eighteenth-Century Algiers: Genealogy and Heritage," *History of the Family* 16, no. 2 (2011): 98–107; and Meropi Anastassiadou, "Des Défunts hors du commun: Les possesseurs de livres dans les inventaires après décès musulmans de Salonique," *Turcica* 32 (2000): 197–252.

25. Ago, *Gusto for Things*, 39.

26. Ago, *Gusto for Things*, xxxii–xxxiii, 93, 216.

27. Elizabeth B. Frierson, "Cheap and Easy: The Creation of Consumer Culture in Late Ottoman Society," in Quataert, *Consumption Studies*, 246; and Toufoul Abou-Hodeib, *A Taste for Home: The Modern Middle Class in Ottoman Beirut* (Stanford, CA: Stanford University Press, 2017), 28, 39, 45, 136.

28. Béchir Yazidi, *La Politique coloniale et le domaine de l'État en Tunisie: De 1881 jusqu'à la crise des années 30* (Tunis: Sahar, 2005), 34, 38.

29. Bargaoui Grangaud, and Cerutti, "Introduction," para.19.

30. Halil İnalcık and Donald Quataert, *An Economic and Social History of the Ottoman Empire, 1300–1914* (Cambridge: Cambridge University Press, 1994), 103; and Kenneth M. Cuno, *The Pasha's Peasants: Land, Society, and Economy in Lower Egypt, 1740–1858* (Cambridge: Cambridge University Press, 1992), 21.

31. Nicolas Michel, *L'Égypte des villages autour du seizième siècle* (Louvain: Peeters, 2018), 174.

32. İnalcık and Quataert, *An Economic and Social History*, 103; and Keith David Watenpaugh, *Being Modern in the Middle East: Revolution, Nationalism, Colonialism, and the Arab Middle Class* (Princeton, NJ: Princeton University Press, 2006), 24.

33. James McDougall, *A History of Algeria* (Cambridge: Cambridge University Press, 2017), 19.

34. McDougall, *A History of Algeria*, 19. See also John Ruedy, *Land Policy in Colonial Algeria* (Berkeley: University of California Press, 1967).

35. McDougall, *A History of Algeria*, 94.

36. McDougall, *A History of Algeria*, 96.

37. Michel, *L'Égypte des villages*; Cuno, *The Pasha's Peasants*, 189–97; Kemal H. Karpat, "The Land Regime, Social Structure, and Modernization in the Ottoman Empire," in *Studies on Ottoman Social and Political History: Selected Articles and Essays* (Leiden: Brill, 2002), 350.

38. Watenpaugh, *Being Modern in the Middle East*, 24.

39. Watenpaugh, *Being Modern in the Middle East*, 357, 375.

40. Michel, *L'Égypte des villages*, 180.

41. Cuno, *The Pasha's Peasants*, 189; Karpat, "The Land Regime," 346.

42. Cuno, *The Pasha's Peasants*, 189, 193.

43. Lisa Anderson, "Nineteenth-Century Reform in Ottoman Libya," *International Journal of Middle East Studies* 16, no. 3 (1984): 332.

44. Anderson, "Nineteenth-Century Reform in Ottoman Libya," 332.

45. Kenneth J. Perkins, *A History of Modern Tunisia* (New York: Cambridge University Press, 2004), 18; Fatma Ben Slimane, "Une 'dhimma inversée?' La question des protections dans la Régence ottomane de Tunis," in *Les musulmans dans l'histoire de l'Europe: II. Passages et contactes en Méditerranée*, ed. Jocelyne Dakhlia and Wolfgang Kaiser (Paris: Albin Michel, 2013), 345–69.

46. Daniela Melfa, *Les Apports culturels des réfugiés politiques italiens en Tunisie au XIXe siècle* (Rome: Aracne, 2008), 117; Julia Ann Clancy-Smith, *Mediterraneans: North Africa and Europe in an Age of Migration, c. 1800–1900* (Berkeley: University of California Press, 2011), 103; J. C. Hurewitz, *The Middle East and North Africa in World Politics: A Documentary Record* (New Haven, CT: Yale University Press, 1975), 1:353; Perkins, *A History of Modern Tunisia*, 22.

47. Gilsenan, "Translating Colonial Fortunes."

48. Abdelhamid Hénia, *Propriété et stratégies sociales à Tunis, XVIe–XIXe siècles* (Tunis: Université Tunis I, 1999), 421, 423.

49. Jacques Revault, *Palais et demeures de Tunis: XVIIIe et XIXe siècles* (Paris: Éditions du Centre national de la recherche scientifique, 1971), 229–32.

50. Husayn Ibn 'Abdallah, Letter 15, April 24, 1867, in *Rasāʾil*, 1:47.

51. TNA, C. 11, d. 102, doc. 7721, Léon Elmilik, *Gouvernement tunisien et succession Heussein (notes explicatives)* (Tunis: Imprimerie internationale, 1889), 72–85.

52. Perkins, *A History of Modern Tunisia*, 20, 25.

53. TNA, C. 11, d. 98, doc. 95, Husayn's deed, February 17, 1861; and d. 113, doc. 8941, French consul in Florence to résident général Massicault, December 30, 1887.

54. Charles C. Harber, "Tunisian Land Tenure in the Early French Protectorate," *Muslim World* 63, no. 4 (1973): 307–15.

55. TNA, C. 11, d. 98, doc. 158, Bey's decree approving the new urban project, June 23, 1863.

56. The *inzāl* disappeared from colonized Algeria in 1844 but was still in use in Tunis under the French protectorate. Jean-Philippe Bras, "L'Imperfection de la propriété indigène, lieu commun de la doctrine juridique coloniale en Afrique du Nord," in Bargaoui Grangaud, and Cerutti, *Appartenance locale*, para. 27. See also Hénia, *Propriété et stratégies sociales à Tunis*, 108.

57. TNA, C. 11, d. 98, doc. 103, David Santillana's report, July 23, 1889; and doc. 158, Beylical decree, June 23, 1863.

58. MAE (Courneuve), CPC, file 630, Husayn's legacy, Orsolina Cubisol to Husayn Ibn 'Abdallah, August 7, 1869.

59. TNA, C. 11, d. 113, doc. 8918, David Santillana to Eugène Régnault, Florence, November 30, 1887; and d. 98, doc. 158, Beylical decree, June 23, 1863.

60. TNA, C. 11, d. 102, doc. 7721, Elmilik, *Gouvernement tunisien et succession Heussein*, 111.

61. TNA, C. 11, d. 102, doc. 7721, 89–90.

62. MAE (Nantes), Protectorat Tunisie, Résidence générale, 1TU/1V/1477, Husayn's legacy (1889), doc. 2.

63. MAE (Courneuve), CPC, file 630, Husayn's legacy, Orsolina Cubisol to Husayn Ibn 'Abdallah, August 7, 1869.

64. TNA, C. 11, d. 98, doc. 24, British Consul Richard Wood to Muhammad al-Sadiq Basha Bey, Tunis, December 19, 1866; and doc. 24, British Consul Richard Wood to Muhammad al-Sadiq Basha Bey, Tunis, January 18, 1867.

65. Perkins, *A History of Modern Tunisia*, 28.

66. Perkins, *A History of Modern Tunisia*, 28.

67. Perkins, *A History of Modern Tunisa*, 40.

68. Malak Labib, "Crise de la dette publique et missions financières européennes en Égypte, 1878–1879," *Monde(s)* 4 (2013): 23–43.

69. M. Şükrü Hanioğlu, *A Brief History of the Late Ottoman Empire* (Princeton, NJ: Princeton University Press, 2008), 136–37; Donald Quataert, *The Ottoman Empire, 1700–1922* (Cambridge: Cambridge University Press, 2005); Christopher Clay, "The Origins of Modern Banking in the Levant: The Branch Network of the Imperial Ottoman Bank, 1890–1914," *International Journal of Middle East Studies* 26, no. 4 (1994): 589–614.

70. Husayn Ibn 'Abdallah, Letter 131, December 24, 1875, in *Rasāʾil*, 2:77.

71. TNA, C. 11, d. 102, doc. 7721, Elmilik, *Gouvernement tunisien et succession Heussein*, 74–75.

72. Al-Shadili b. Salih was an Islamic scholar "born about 1810 in Tunis. . . . In 1873 he was elevated to the position of Bash Mufti" al-Maliki. Arnold H. Green, *The Tunisian Ulama 1873–1915: Social Structure and Response to Ideological Currents* (Leiden: Brill, 1978), 263.

73. TNA, C. 11, d. 102, doc. 7721, Elmilik, *Gouvernement tunisien et succession Heussein*, 74; and d. 98, doc. 103, David Santillana's report, July 23, 1889.

74. Green, *The Tunisian Ulama*, 148, 287, 288. Ahmad al Wartatani was an Islamic scholar born in 1831 near Le Kef. He served as "vice-president in the *Awqāf* Administration from 1874 until 1878" before becoming the president of this administration (Green, *The Tunisian Ulama*, 288). He passed away in 1885.

75. TNA, C. 11, d. 102, doc. 7721, Elmilik, *Gouvernement tunisien et succession Heussein*, 75–76.

76. TNA, C. 11, d. 98, doc. 103, David Santillana's report, July 23, 1889.

77. Husayn Ibn 'Abdallah, Letter 31, June–July 1869, in *Rasāʾil*, 1:94.

78. Revault, *Palais et demeures de Tunis*, 229–32.

79. Husayn Ibn 'Abdallah, Letter 204, February 28, 1879, in *Rasāʾil*, 3:16; and Letter 229, January 7, 1880, in *Rasāʾil*, 3:97.

80. Husayn Ibn 'Abdallah, Letter 240, August 2, 1880, in *Rasāʾil*, 3:91–92.

81. Husayn Ibn 'Abdallah, Letter 242, September 1, 1880, in *Rasāʾil*, 3:95.

82. Colette Establet, *Familles et fortunes à Damas: 450 foyers damascains en 1700* (Damas: Institut français de Damas, 1994); Pascale Ghazaleh, *Held in Trust: Waqf in the Islamic World* (Cairo: American University in Cairo Press, 2011), 150; Didier Guignard, *L'Abus de pouvoir dans l'Algérie coloniale, 1880–1914* (Paris: Presses universitaires de Paris Ouest, 2010). After Tunisia's declaration of bankruptcy in 1867, rural and urban communities alike sank into debt. See Hénia, *Propriété et stratégies sociales à*

Tunis, 369; and Sami Bargaoui, "Le Phénomène de l'usure dans la Tunisie de 1881," in *Réactions à l'occupation française de la Tunisie en 1881: Actes du 1er Séminaire d'histoire du Mouvement National: 29, 30 et 31 mai 1981, Hôtel Sidi Dhrif, Sidi Bou Saïd, Tunisie* (Tunis: Tunis University, 1983), 353–57.

83. TNA, C. 11, d. 97, doc. 33, translation into Arabic of a letter sent by Giaccomo Guttierez, April 6, 1882; Biblioteca Labronica, Livorno, Husayn Ibn 'Abdallah, *In nome di Sua Maestà Umberto I* (1886), 26–27; TNA, C. 11, d. 103, doc. 7729, al-Hadi Zarruq to Prime Minister Muhammad al-'Aziz Bu 'Attur, Cairo, July 17, 1887.

84. TNA, C. 11, d. 97, doc. 45, Agreement between General Husayn and Giaccomo Guttierez; see also ibid., TNA, C. 11, d. 98, doc. 41, Finance director to Tunisian prime minister, Tunis, May 21, 1891: "il a été stipulé notamment qu'une somme de 200 000 francs due à Hussein serait retenu durant cinq ans en prévision des répétitions à exercer par le gouvernement. Cette somme devait produire intérêt à 5% au profit de Hussein."

85. TNA, C. 11, d. 103, doc. 7725, Contract between the French Army and Husayn Ibn 'Abdallah represented by Léon Elmilik, April 8, 1882; TNA, C. 11, d. 103, doc. 7723, Beylical decree, April 6, 1890; Revault, *Palais et demeures de Tunis*, 229–32.

86. Husayn Ibn 'Abdallah, Letter 288, August 9, 1886, in *Rasāʾil*, 3:188.

87. TNA, C. 11, d. 99, 'Amr Bu Hajib, June 18, 1891, f. 17; and Husayn, Letter 288, in *Rasāʾil*, 3:188.

88. Byron Cannon, "Le Marché de location des habous en Tunisie: Dialectique de développement agricole," in *Terroirs et Sociétés au Maghreb et Moyen Orient: Séminaire IRMAC, 1983–84, Table ronde franco-américaine CNRS/NSF, Lyon, juin 1984* (Lyon: Maison de l'Orient, 1987), 97.

89. Harber, "Tunisian Land Tenure," 311.

90. Harber, "Tunisian Land Tenure," 312.

91. Cannon, "Le Marché de location des habous en Tunisie," 98.

92. McDougall, *A History of Algeria*, 98.

93. McDougall, *A History of Algeria*, 142.

94. Yazidi, *La Politique coloniale et le domaine de l'Etat*.

95. For a very different context, see Kenneth Pommeranz, *The Great Divergence: China, Europe and the Making of the Modern World Economy* (Princeton, NJ: Princeton University Press, 2000).

96. Jocelyne Dakhlia and Bernard Vincent, eds., *Les Musulmans dans l'histoire de l'Europe* (Paris: Albin Michel, 2011).

3. A World of "Affairs"

1. Jean Ganiage, *Les Origines du Protectorat français en Tunisie, 1861–1881* (Paris: Presses universitaires de France, 1959), 161–62; and Leon Carl Brown, *The Tunisia of Ahmad Bey, 1837–1855* (Princeton, NJ: Princeton University Press, 1974).

2. Tunisian National Archives (TNA), C. 11, d. 97, doc. 52, David Santillana to Husayn Ibn 'Abdallah, Tunis, March 25, 1884.

3. Lauren Benton, "Colonial Law and Cultural Difference: Jurisdictional Politics and the Formation of the Colonial State," *Comparative Studies in Society and History* 41, no. 3 (1999): 563–88; Mary Dewhurst Lewis, *Divided Rule: Sovereignty and Empire in French Tunisia, 1881–1938* (Berkeley: University of California Press, 2013); Paolo Sartori, "Premessa. Waqf, colonialismo e pluralismo giuridico nelle società musulmane," *Quaderni storici* 132 (2009): 627–52.

4. Romain Bertrand, "Politiques du moment colonial. Historicités indigènes et rapports vernaculaires au politique en 'situation coloniale,'" *Questions de recherche* 26 (2008): 1–49; and Jean-François Bayart, "Les Études postcoloniales, une invention politique de la tradition?," *Sociétés politiques comparées. Revue européenne d'analyse des sociétés politiques* 14 (2009): 1–46.

5. Ismail Warscheid, *Traduire le social en normatif: La justice islamique dans le grand Touat (Sahara algérien) au XVIIIe siècle* (Paris: EHESS, 2014).

6. Lewis, *Divided Rule*, xiii: "This is a book about a place where the majority of people spoke Arabic but at a time when very few of them wrote it. Accordingly, only a few of my sources were in Arabic."

7. Ministère des Affaires étrangères (MAE) (Courneuve), CPC, file 602, Ben Ayed's Affair, Note on Ben Ayed, May 13, 1913.

8. Ganiage, *Les Origines du Protectorat français*, 161; Brown, *The Tunisia of Ahmad Bey*.

9. MAE (Courneuve), CPC, file 602, Ben Ayed's Affair, Note on Ben Ayed, May 13, 1913.

10. Abdelhamid Larguèche, "Nasîm Shammamâ: Un caïd face à lui-même et face aux autres," in *Juifs et Musulmans en Tunisie: Fraternité et déchirements*, ed. Sonia Fellous (Paris: Somogy, 2003), 146.

11. Larguèche, "Nasîm Shammamâ," 146; Husayn Ibn 'Abdallah, Letter 32, August 10, 1869, in *Rasā'il Ḥusayn ilā Khayr al-Dīn*, ed. Aḥmad 'Abd al-Sālam (Carthage: Bayt al-Ḥikma, 1991–92), 1:99; Biblioteca Labronica, Livorno (BL), Michel Ersilio file, *Lettera del generale Heussein agli onorevoli avvocati componenti il collegio della difesa del governo di Tunis (traduzione dall'arabo)* (Florence: Ricci, 1880), 25.

12. Habib Jamoussi, *Juifs et chrétiens en Tunisie au 19e siècle. Essai d'une étude socioculturelle des communautés non-musulmanes (1815–1881)* (Sfax: Amal Éditions, 2010), 281; and Ganiage, *Les Origines du Protectorat français*, 170.

13. MAE (Courneuve), CPC, file 602, Ben Ayed's Affair, Note on Ben Ayed, May 13, 1913.

14. Yavel Harouvi, "Les Conflits autour du testament du Caïd Nessim Scemama d'après quelques sources hébraïques," in *Entre Orient et Occident: Juifs et Musulmans en Tunisie*, ed. Denis Cohen-Tannoudji (Paris: Éditions de l'éclat, 2007), 143–56.

15. Harouvi, "Les Conflits autour du testament," 144–47.

16. Luc Boltanski, Elisabeth Claverie, Nicolas Offenstadt, and Stéphane Van Damme, eds., *Affaires, scandales et grandes causes: De Socrate à Pinochet* (Paris: Stock, 2007), 227.

17. Rome, Archivio storico, Ministero degli Affari Esteri, Serie Politica, Tunisi, n°522, rapporto in arrivo 1440 (198), December 30, 1871–December 28, 1875, Consul Pinna to Italian Ministry of Foreign Affairs, Visconti Venosta, May 7, 1872.

18. Ganiage, *Les Origines du Protectorat français*, 601.

19. MAE (Nantes), 1TU/1V/1211A, Tunisian Protectorate, Caïd Eliaou Shemama's files.

20. MAE (Nantes), 1TU/1V/314 (F-L), 3, Tunisian Protectorate, Public Record office PRO, Foreign Office 335/166, Correspondence Algiers-Tunis, 1886–1887, Playfair to Foreign Office, March 9, 1887; Hamida ben Ayed, letter from the house of the representatives, chamber of deputies, Paris, April 30, 1888; Hamida Ben Ayad's claims to résident général Massicault, Tunis, January 24, 1882.

21. Jocelyne Dakhlia and Bernard Vincent, eds., *Les Musulmans dans l'histoire de l'Europe* (Paris: Albin Michel, 2011); and Nabil Matar, *Islam in Britain, 1558–1685* (Cambridge: Cambridge University Press, 1998).

22. Robert Attal, *Le Caïd Nessim Samama de Tunis: Mécène du livre hébraïque* (Jerusalem: Attal, 1995); and Gilles Boulu, "Recherches sur les Scemama ou Samama de Tunis. Une dynastie de personnalités et de hauts-fonctionnaires de la régence de Tunis" (unpublished manuscript available at https://docplayer.fr/17617165-Recherches-sur-les-scemama-ou-samama-de-tunis.html).

23. Didier Guignard, *L'Abus de pouvoir dans l'Algérie coloniale, 1880–1914* (Paris: Presses universitaires de Paris Ouest, 2010), 27.

24. Pierre-Cyrille Hautcœur, "Les Transformations du crédit en France au XIXe siècle," *Romantisme* 151, no. 1 (2011): 23–38.

25. Boltanski et al., *Affaires, scandales et grandes causes*, 12, 397–98, 412.

26. Boltanski et al., *Affaires, scandales et grandes causes*, 414.

27. Rome, Archivio storico, Ministero degli Affari Esteri, Serie Politica, Tunisi, n°522, rapporto in arrivo 1440 (198), December 30, 1871–December 28, 1875, Consul Pinna to Italian Ministry of Foreign Affairs, Visconti Venosta, March 4, 1873.

28. Ganiage, *Les Origines du Protectorat français*; and Guignard, *L'Abus de pouvoir*, 89.

29. TNA, C. 11, d. 112, doc. 8738, Agreement on the Scemama affair with General Muhammad Bakkush, Livorno, July 6, 1884; and Ganiage, *Les Origines du Protectorat français*, 210, 288, 308–9.

30. Husayn Ibn 'Abdallah, Letter 250, October 13, 1881, in *Rasā'il*, 3:108.

31. Husayn Ibn 'Abdallah, Letter 268, August 22, 1884, in *Rasā'il*, 3:145.

32. MAE (Courneuve), CPC, file 630, Husayn's legacy, Husayn's papers. Other words such as *qaḍiyya* and *masʾala* are also used in Arabic.

33. Husayn Ibn 'Abdallah, Letter 28, September 30, 1868, in *Rasā'il*, 1:87. Ibn 'Ayyad received letters from Ahmad b. Abi al-Dhiyaf, principal secretary to the *beys* of Tunis and the most important chronicler of the *beys*' dynastic history in the nineteenth century.

34. Harouvi, "Les Conflits autour du testament," 144–47.

35. Husayn Ibn 'Abdallah, Letter 100, April 11, 1874, in *Rasā'il*, 2:11.

36. TNA, C. 11, d. 113b, doc. 4011–4036, 1873–Feb. 1874, Notes on Husayn's expanses in handling caïd Nessim's case.

37. TNA, C. 11, d. 109, doc. 7989, Husayn Ibn 'Abdallah to Prime Minister Si Muhammad Khaznadar; TNA, C. 11, d. 105, doc. 7730, Husayn Ibn 'Abdallah to Mustafâ Murâlî, June 6, 1882; Husayn Ibn 'Abdallah, Letter 90, September 6, 1873, in *Rasā'il*, 1:201; BL, *Lettera del generale Heussein*, 25.

38. BL, *In nome di sua Maesta Umberto 1* (1886), 1.

39. TNA, C. 11, d. 102, doc. 7721, Léon Elmilik, *Gouvernement tunisien et succession Heussein (notes explicatives)* (Tunis: Imprimerie internationale, 1889), 40–45.

40. TNA, C. 11, d. 105, doc. 7767, Note on Moraly's case, David Santillana, October 10, 1887.

41. TNA, C. 11, d. 102, doc. 7721, Elmilik, *Gouvernement tunisien et succession Heussein*, 40–45.

42. TNA, C. 11, d. 105, doc. 7773, Husayn Ibn 'Abdallah to Mustafa al-Murali, May 2, 1878.

43. TNA, C. 11, d. 105, doc. 7777, Husayn Ibn 'Abdallah to Hay al-Sabbagh, confidential letter (*sarrî*), May 1878.

44. TNA, C. 11, d. 105, doc. 7781, Husayn Ibn 'Abdallah to Sabbagh's sons, February 5, 1885.

45. TNA, C. 11, d. 105, doc. 7767, Note on Moraly's case, David Santillana, October 10, 1887.

46. TNA, C. 11, d. 102, doc. 7721, Elmilik, *Gouvernement tunisien et succession Heussein*, 40–45; TNA, C. 11, d. 105, doc. 7786 and 7786bis, Giaccomo Guttierez to Bu 'Attur, Livorno, August 19, 1884.

47. TNA, C. 11, d. 105, doc. 7738, Translation of Judeo-Arabic text into Italian by Elyahu and Joseph Sabbagh, Livorno, November 30, 1884.

48. TNA, C. 11, d. 102, doc. 7721, Elmilik, *Gouvernement tunisien et succession Heussein*, 40–45.

49. TNA, C. 11, d. 102, doc. 7721; TNA, C. 11, d. 102, doc. 19, Copy of the agreement between Husayn and Tunisian government in 1884, March 22, 1892.

50. TNA, C. 11, d. 109, doc. 7997, Husayn Ibn 'Abdallah to Prime Minister Mustafa b. Isma'il, November 10, 1879.

51. TNA, C. 11, d. 109, doc. 8304, Note on Santoni di Sio's case, David Santillana, April 21, 1887; TNA, C. 11, d. 109, doc. 8047, Husayn Ibn 'Abdallah to Prime Minister Mustafa b. Isma'il, December 19, 1879.

52. TNA, C. 11, d. 109, doc. 8047, Husayn Ibn 'Abdallah to Prime Minister Mustafa b. Isma'il, December 19, 1879.

53. TNA, C. 11, d. 102, doc. 19, Copy of the agreement between Husayn and Tunisian government in 1884, March 22, 1892; TNA, C. 11, d. 109, doc. 8304, Note on Santoni di Sio's case, David Santillana, April 21, 1887; Guignard, *L'Abus de pouvoir*, 15, 24–25; John C. Rayburn, "Count Joseph Telfener and the New York, Texas, and Mexican Railway Company," *Southwestern Historical Quarterly* 68, no. 1 (1964): 29–42.

54. TNA, C. 11, d. 109, doc. 8331, Scemama's legacy, December 10, 1889; TNA, C. 11, d. 109, doc. 8304, Note on Santoni di Sio's case, David Santillana, April 21, 1887.

55. TNA, C. 11, d. 109, doc. 8331, Scemama's legacy, December 10, 1889.

56. TNA, C. 11, d. 102, doc. 7721, Elmilik, *Gouvernement tunisien et succession Heussein*, 28; TNA, C. 11, d. 109, doc. 8297, Note on Santoni di Sio's case, David Santillana, October 29, 1887.

57. Husayn Ibn 'Abdallah, Letter 193, October 11, 1878, in *Rasāʾil*, 2:185.

58. Marc Aymes, "Agence par défaut," in *L'Art de l'État en Turquie: Arrangements de l'action publique de la fin de l'Empire ottoman à nos jours* (Paris: Karthala, 2013), 35–52.

59. Guillaume Calafat, "Une Mer jalousée: Juridictions maritimes, ports francs et régulation du commerce en Méditerranée (1590–1740)" (Ph.D. diss., Paris 1 Panthéon -Sorbonne, 2013), 538.

60. Calafat, "Une Mer jalousée," 401; see also Lucette Valensi, *Le Maghreb avant la prise d'Alger, 1790–1830* (Paris: Flammarion, 1969).

61. Jean-Pierre Filippini, "Les Juifs d'Afrique du Nord et la communauté de Livourne au XVIIIe siècle," in *Les Relations intercommunautaires juives en Méditerranée occidentale XIIIe-XXe siècle*, ed. Jean-Louis Miège (Paris: CNRS Editions, 1984), 50–63.

62. TNA, C. 11, d. 97, doc. 84, David Santillana to Giaccomo Guttierez, Livorno, September 16, 1883.

63. MAE (Courneuve), CPC, file 629, Husayn's legacy, Agreement with Léon Elmilik, Paris; Silvia Falconieri, "David Santillana," in *Dictionnaire des juristes ultramarins XVIIIe-XXe siècles*, ed. Florence Renucci (Paris: Mission de recherche Droit et justice, 2012), 272–76; Anna Baldinetti, *David Santillana, l'uomo e il giurista (1855–1931): scritti inediti, 1878–1920* (Rome: Istituto per l'Oriente C.A. Nallino, 1995); Florence Renucci, "David Santillana, acteur et penseur des droits musulman et européen" *Monde(s)* 7, no. 1 (2015): 25–44.

64. TNA, C. 11, d. 113, doc. 8849, French consul in Florence to French Ministry of Foreign Affairs, October 16, 1887; TNA, C. 11, d. 113b, doc. 8967, *Bertucci and Keusch vs. Elmilik, Paoli and de Laigue*, Florence, 1887, 15.

65. TNA, C. 11, d. 105, doc. 7796, Giaccomo Guttierez to Régnault, Livorno, February 23, 1887; MAE (Nantes), 22 PO/1/100bis, October–November 1886, Husayn's legacy, two letters from Hay Sabbagh, April 10, 1877; Rome, Archivio storico, Ministero degli Affari Esteri, Serie Politica, Tunisi, n°522, rapporto in arrivo 1440 (198), December 30, 1871–December 28, 1875, Consul Pinna to Italian Ministry of Foreign Affairs, Visconti Venosta, May 7, 1872.

66. MAE (Courneuve), CPC, file 630, Husayn's legacy, Envelope 49.

67. Ganiage, *Les Origines du Protectorat français*, 478. Muhammad al-Bakkūsh (1833–96) was married to a woman from the Ibn 'Ayyad family. In 1860, he became the first secretary to Prime Minister Khaznadar and then the head of the Department of Foreign Affairs. He was involved in negotiating a loan with Erlanger Bank.

68. MAE (Courneuve), CPC, file 630, Husayn's legacy, Envelope 49, Rosa.

69. BL, *S.E. il generale Heussein contro Salomone Cohen Rosa* (Livorno: Zecchini, 1881), 68–69; and BL, *In nome di sua Maesta Umberto 1*, 40.

70. Filippini, "Les Juifs d'Afrique du Nord," 63; Linda Funaro, "A Mediterranean Diaspora: Jews from Leghorn in the Second Half of the 19th Century," in *L'Europe méditerranéenne/Mediterranean Europe*, ed. Marta Petricioli (Brussels: Peter Lang, 2008), 100.

71. TNA, C. 11, d. 113b, doc. 9233, Agreement between Léon Elmilik and David Santillana, Paris, July 1, 1888.

72. Lucca, Archivio di Stato, Corte di appello di Lucca, 22–1887, March 29, 1887, "Generale Heussein vs. Elmilik Leone," 36.

73. TNA, C. 11, d. 103, doc. 7725, Contract between the French Army and Husayn Ibn 'Abdallah represented by Léon Elmilik, April 8, 1882.

74. BL, *S.E. il generale Heussein contro Salomone Cohen Rosa*, 3, 13.

75. Husayn, Letter 250, October 13, 1881, in *Rasā'il*, 3:108–10; TNA, C. 11, d. 113, doc. 8788, Copies of two reports, French consul in Florence to Massicault, Paris, July 17, 1887.

76. Husayn Ibn 'Abdallah, Letter 265, December 6, 1883, in *Rasā'il*, 3:140; Letter 268, August 22, 1884, in *Rasā'il*, 3:145; MAE (Nantes), 227 PO/1/100bis, October–November 1887, Husayn's legacy.

77. TNA, C. 11, d. 112, doc. 8738, Agreement on Scemama's case including with Bakkush, Livorno, July 6, 1884; TNA, C. 11, d. 105, doc. 7788, Léon Elmilik, credits and rights' assignments, Livorno, December 14, 1884.

78. MAE (Nantes), 227 PO/1/100bis, October–November 1887, Husayn's legacy, 'Amr Bu Hajib to Husayn Ibn 'Abdallah, f. 35 and 78.

79. TNA, C. 11, d. 113b, doc. 9153, David Santillana to Secretary General Régnault, Florence, May 4, 1888.

80. TNA, C. 11, d. 98, doc. 32, Statement by Attia Guetta, Livorno, June 5, 1890.

81. MAE (Nantes), 227 PO/1/100bis, October–November 1887, Husayn's legacy, Anonymous to French consul, Livorno, October 24, 1887.

82. TNA, C. 11, d. 109, doc. 8047, Husayn Ibn 'Abdallah to Prime Minister Mustafa b. Isma'il, December 19, 1879.

83. Warscheid, *Traduire le social en normatif*, 268, 271. See also Sami Bargaoui, "Les Titres fonciers dans la régence de Tunis à l'époque moderne: Interrogations autour d'une mutation documentaire," *IBLA* 208 (2011): 168–69, 180.

84. Johansen, "Formes de langage et fonctions publiques," 337, 361; Boğaā A. Ergene, "Evidence in Ottoman Courts: Oral and Written Documentation in Early-Modern Courts of Islamic Law," *Journal of the American Oriental Society* 124, no. 3 (2004): 471–91.

85. Johansen, "Formes de langage et fonctions publiques," 276; Brinkley Messick, *The Calligraphic State: Textual Domination and History in a Muslim Society* (Berkeley: University of California Press, 1993), 217.

86. Johansen, "Formes de langage et fonctions publiques," 337.

87. Messick, *The Calligraphic State*, 217.

88. Michael Gilsenan, "Possessed of Documents: Hybrid Laws and Translated Texts in the Hadhrami Diaspora," *Comparative Studies of South Asia, Africa and the Middle East* 31, no. 2 (2011): 368–69.

89. Isabelle Grangaud, "Prouver par l'écriture: Propriétaires algérois, conquérants français et historiens ottomanistes," *Genèses* 74, no. 1 (2009): 36–39.

90. Johansen, "Formes de langage et fonctions publiques"; Robert Brunschvig, *Études d'islamologie* (Paris: Maisonneuve et Larose, 1976), 215; Ergene, "Evidence in Ottoman Courts," 477; Rachida Chih, Catherine Mayeur-Jaouen, and Rüdiger Seesemann, "The Nineteenth Century: A Sufi Century?," in *Sufism, Literary Production, and Printing in the Nineteenth Century* (Würzburg: Ergon Verlag, 2015), 8; Messick, *The Calligraphic State*, 54–55.

91. Brinkley Messick, "Evidence: From Memory to Archive," *Islamic Law and Society* 9, no. 2 (2002): 268–69.

92. Manuel Charpy and François Jarrige, "Introduction. Penser le quotidien des techniques. Pratiques sociales, ordres et désordres techniques au XIXe siècle," *Revue d'histoire du XIXe siècle* 45 (2012): 25.

93. Brunschvig, *Études d'islamologie*, 212; Messick, "Evidence," 231, 255; Emile Tyan, "'Adl," *Encyclopaedia of Islam*, 2nd ed., http://dx.doi.org/10.1163/1573-3912_islam_COM_0019; Khaled Fahmy, *In Quest of Justice. Islamic Law and Forensic Medicine in Modern Egypt* (Oakland: University of California Press, 2018), 97.

94. Messick, *The Calligraphic State*, 253.

95. Chih, Mayeur-Jaouen, and Seesemann, "The Nineteenth Century," 17.

96. Marc Aymes, "Changeur d'empire," in *Penser, agir et vivre dans l'empire ottoman et en Turquie: Études réunies pour François Georgeon*, ed. François Georgeon, Nathalie Clayer, and Erdal Kaynar (Paris: Peeters, 2013), 269.

97. Guignard, *L'Abus de pouvoir*, 308.

98. TNA, C. 11, d. 97, doc. 16, Letter to Giaccomo Guttierez.

99. Isabelle Grangaud, "Affrontarsi in archivio tra storia ottomana e storia coloniale (algeri 1830)," *Quaderni storici* 129, no. 3 (2008): 641.

100. TNA, C. 11, d. 109, doc. 8023, Husayn Ibn 'Abdallah to Prime Minister Mustafa b. Isma'il, February 28, 1880.

101. TNA, C. 11, d. 113, doc. 8811, De Laigue to résident général Massicault, August 19, 1887.

102. TNA, C. 11, d. 113, doc. 8762, Note from David Santillana, Florence, July 1, 1887.

103. TNA, C. 11, d. 97, doc. 75, Giaccomo Guttierez to Prime Minister Sidi Mustafa, Livorno, July 1, 1881.

104. Rome, Archivio storico, Ministero degli Affari Esteri, II. registri copia lettere in partenza: Dispacci inviati dal ministero alle varie legazioni e consolati, Tunisi, 1223 (127), February 10, 1871–June 7, 1879, Royal Department to Husayn Ibn 'Abdallah, September 12, 1876.

4. The Diplomatic Conflicts Over Husayn's Estate

1. Ministère des Affaires étrangères (MAE), Nantes, 227 PO/1/100, Husayn's file (1887–88), Léon Elmilik to Consul, Livorno, July 8, 1887.

2. Tunisian National Archives (TNA), C. 11, d. 113, doc. 8850, David Santillana to Régnault, Florence, October 16, 1887.

3. TNA, C. 11, d. 113, doc. 8764, French Consul to Ottoman Consul, Florence, July 1, 1887; and doc. 8765, French Consul to Ottoman Consul, July 2, 1887.

4. Olivier Bouquet and Sinan Kuneralp, *Les Musurus, une famille de diplomates ottomans: Lettres et documents (1852–1920)* (Istanbul: Isis, 2015), 14, 283–85, 307–9, 312–13; Basile Musurus Ghikis to Etienne Musurus, letter 174, May 16, 1885; Basile Musurus Ghikis to Constantin Musurus, letter 191, March 27, 1886; Basile Musurus Ghikis to Constantin Musurus, letter 195, January 3/13, 1887.

5. TNA, C. 11, d. 113, doc. 8763, French Consul de Laigue to French Foreign Office; and doc. 8819, French Consul de Laigue to French Foreign Office, September 7, 1887.

6. Başbakanlık Osmanlı Arşivi (BOA), HR.H . . . 171/13, Argument of the legal advisers in Istanbul, July 11, 1887 (trans. Nilüfer Hatemi); and Cable from Saïd Pacha, Minister of Ottoman Foreign Affairs, to Essada Pacha, Ottoman Ambassador in Paris,

Paris, July, 6, 1887. About Essada (or Esat) Pacha, Ottoman ambassador in Paris from 1880 to 1894, see Sinan Kuneralp, "Diplomates et consuls ottomans en France au XIXe siècle," in *L'Empire Ottoman, la République de Turquie et la France*, ed. Hâmit Batu and J.-L. Bacqué-Grammont (Istanbul: Isis, 1996), 307.

7. Abdurrahman Çaycı, *La Question tunisienne et la politique ottomane, 1881–1913* (Ankara: Imprimerie de la Société Turque d'Histoire, 1992), 160.

8. TNA, C. 11, d. 113, doc. 8820, French Consul to French Foreign Office, Florence, September 7, 1887.

9. Kenneth Perkins, *A History of Modern Tunisia* (New York: Cambridge University Press, 2014), 37; and Augustin Bernard, "Le Recensement de 1906 en Algérie et en Tunisie," *Annales de Géographie* 17, no. 91 (1908): 30.

10. Gabriele Montalbano, "Les Italiens de Tunisie: La construction de l'italianité dans un contexte colonial français (1896–1918)" (Ph.D. diss., École Pratique des Hautes Études, 2018), 183.

11. Montalbano, "Les Italiens de Tunisie," 7.

12. Hachemi Karoui and Ali Mahjoubi, *Quand le soleil s'est levé à l'Ouest. Tunisie 1881: Impérialisme et Résistance* (Tunis: Cérès, 1983), 136–43; and Perkins, *A History of Modern Tunisia*, 42.

13. William I. Shorrock, "The Tunisian Question in French Policy Toward Italy, 1881–1940," *International Journal of African Historical Studies* 16, no. 4 (1983): 633.

14. See Çaycı, *La Question tunisienne*; and Kenneth Perkins, " 'The Masses Look Ardently to Istanbul,' " in *Islamism and Secularism in North Africa*, ed. John Ruedy (New York: St. Martin's Press, 1994), 23–36.

15. Andreas Tunger-Zanetti, *La Communication entre Tunis et Istanbul, 1860–1913: Province et métropole* (Paris: L'Harmattan, 1996).

16. Juliette Bessis, *La Méditerranée fasciste. L'Italie mussolinienne et la Tunisie* (Paris: Karthala, 1981); Mary Dewhurst Lewis, *Divided Rule: Sovereignty and Empire in French Tunisia, 1881–1938* (Berkeley: University of California Press, 2014); Julia Ann Clancy-Smith, *Mediterraneans: North Africa and Europe in an Age of Migration, c. 1800–1900* (Berkeley: University of California Press, 2012); Montalbano, "Les Italiens de Tunisie."

17. For instance, Edouard Clunet, *Incident du consulat à Florence* (Paris: Marchal et Billard, 1888).

18. 'Abd al-Qādir Vaḥḥah, *Al-Ḥaraka al-iṣlāḥīya al-zaytūnīya wa-al-madrasīya al-Tūnisīya: al-Shaykh Sālim Bū Ḥājib wa-manhajuhu al-islāhī 1342–1244 H/1827–1828–1924* ([Tunis]: 'Abd al-Qādir Vahhah, 2007), 358.

19. Perkins, " 'The masses look ardently to Istanbul,' " 26.

20. TNA, C. 11, d. 113, doc. 8819, French Consul to French Foreign Office, Florence, September 6, 1887.

21. TNA, C. 11, d. 113, doc. 8804, French Consul to French Foreign Office, Florence, August 22, 1887; MAE (Nantes), 227 PO/1/100, Husayn's file (1887–88), Inventories de Laigue and Musurus Guikis.

22. Husayn Ibn 'Abdallah, Letter 238, June 9, 1880, in *Rasā'il Ḥusayn ilā Khayr al-Dīn*, ed. Aḥmad 'Abd al-Sālam (Carthage: Bayt al-Ḥikma, 1992), 3: p. 87.

23. Pierre Bardin, *Algériens et Tunisiens dans l'Empire Ottoman de 1848 à 1914* (Paris: CNRS, 1980), 109–10.

24. Thomas Eich, "The Forgotten Salafī-Abū l-Hudā As-Sayyādī," *Die Welt des Islams* 43, no. 1 (2003): 61–87; Butrus Abu-Manneh, "Sultan Abdulhamid II and Shaikh Abulhuda Al-Sayyadi," *Middle Eastern Studies* 15, no. 2 (1979): 131–53; Anne-Laure Dupont, "An Expression of Pan-Islamism in Tunisia at the Beginning of the French Protectorate: The Critique of the Tijāniyya in the *Riḥla al-hijāziyya* by Muhammad b. 'Uthmān al-Sanūsi (d. 1900)" in *Sufism, Literary Production, and Printing in the Nineteenth Century*, ed. Rachida Chih, Cathering Mayeur-Jaouen, and Rüdiger Seesemann (Würzburg: Ergon, 2015), 402; Tunger-Zanetti, *La Communication*, 67.

25. Tunger-Zanetti, *La Communication*, 58.

26. Bardin, *Algériens et Tunisiens*, 21–26; Tunger-Zanetti, *La Communication*, 72–76; Byron D. Cannon and Muhammad Safiullah, "The Function of Wakalah in Intra-Ottoman Relations: Egypt and Tunisia in the First Half of the 19th Century," *Arab Historical Review for Ottoman Studies* 5–6 (1992): 25–29; Mongi Smida, *Consuls et consulats de Tunisie au 19e siècle* (Tunis: Imprimerie de l'Orient, 1991), 40, 48, 50.

27. Tunger-Zanetti, *La Communication*, 174.

28. Vaḥḥah, *Al-Ḥaraka al-iṣlāḥīya al-zaytūnīya wa-al-madrasīya*, 358; TNA, C. 11, d. 99, doc. 3, 'Amr Bu Hajib's reply, June 18, 1891; N. J. Coulson, C. Cahen, B. Lewis, and R. Le Tourneau, "Bayt al-Māl," *Encyclopaedia of Islam*, 2nd ed., accessed February 12, 2019, http://dx.doi.org/10.1163/1573-3912_islam_COM_0109.

29. Robert Mantran, "L'Evolution des relations entre la Tunisie et l'Empire Ottoman du XVIe au XIXe siècle," Cahiers de Tunisie (1959): 58; Tunger-Zanetti, *La Communication*, 4–5; Abdeljelil Temimi, *Recherches et documents d'histoire maghrébine; la Tunisie, l'Algérie et la Tripolitaine de 1816 à 1871* (Tunis: Université de Tunis, 1980); Ali Zouari, *Les Relations commerciales entre Sfax et le Levant aux XVIIIe et XIXe siècles* (Tunis: Institut national d'archéologie et d'art, 1990).

30. TNA, C. 11, d. 113, doc. 9000, French "résident général" to French Foreign Affairs Minister, Tunis, May 25, 1881.

31. Abdurrahman Çaycı, "La Question de la suppression des consulats tunisiens en Italie et l'entente italo-turque de 1863," *Cahiers de Tunisie* (1966): 41–52.

32. TNA, C. 11, d. 113, doc. 8802, David Santillana to French "résident général," Florence, August 14, 1887.

33. Mongi Smida, *Khereddine: Ministre réformateur, 1873–1877* (Tunis: Maison tunisienne de l'édition, 1971); Mohamed-Hédi Cherif, *Pouvoir et société dans la Tunisie de H'usayn bin 'Ali, 1705–1740*, 2 vols (Tunis: Université de Tunis, 1986).

34. BOA, HR.H . . . 171/13, Argument of the legal advisers in Istanbul, July 11, 1887 (trans. Nilüfer Hatemi).

35. BOA, HR.H . . . 171/13, Argument of the legal advisers in Istanbul, July 11, 1887 (trans. Nilüfer Hatemi); and Cable from Yanko Photiades Pacha, Ottoman Ambassador to Rome, to the Minister of Ottoman Foreign Affairs Saïd Pacha, Rome, July 5, 1887. About Yanko Photiades, see Christine Philliou, *Biography of an Empire: Governing Ottomans in an Age of Revolution* (Berkeley: University of California Press, 2011), 233n9.

36. TNA, C. 11, d. 113, doc. 8900, French Consul to French Foreign Office, Florence, November 22, 1887.

37. TNA, C. 11, d. 113, doc. 8767, Cable [no specific author], July 3, 1887.

38. TNA, C. 11, d. 113, doc. 8764, French Consul to Ottoman Consul, Florence, July 1, 1887; and doc. 8785, David Santillana to French Consul, July 12, 1887.

39. TNA, C. 11, d. 113, doc. 8786, David Santillana to Secretary General Régnault, Florence, July 15, 1887.

40. TNA, C. 11, d. 113, doc. 8762, Note written by David Santillana, Florence, July 1, 1887.

41. TNA, C. 11, d. 113, doc. 8786, David Santillana to Secretary General Régnault, Florence, July 15, 1887.

42. TNA, C. 11, d. 113, doc. 8804, French Consul to French Foreign Office, Florence, August 22, 1887; and doc. 8820, French Consul to French Foreign Office, Florence, September 7, 1887.

43. Bardin, *Algériens et Tunisiens*; Noureddine Amara, "Les nationalités d'Amîna Hanım: Une pétition d'hérédité à la France (1830–1896)," *Revue des mondes musulmans et de la Mediterranée* no. 137 (2015): 49–72; Noureddine Amara, "La Nationalité algérienne des Touatis: Un événement à la mesure d'empire (1830–1901)," *Maghreb et sciences sociales* (2012): 99–106; Will Hanley, *Identifying with Nationality: Europeans, Ottomans, and Egyptians in Alexandria* (New York: Columbia University Press, 2017), 189.

44. Hanley, *Identifying with Nationality*, 20, 88.

45. Lewis, *Divided Rule*, 116; Laure Blévis, "La Citoyenneté française au miroir de la colonisation," *Genèses* 53, no. 4 (2003): 25–47; Emmanuelle Saada, *Les Enfants de la colonie: Les métis de l'empire français entre sujétion et citoyenneté* (Paris: Découverte, 2007).

46. M'hamed Oualdi, "La Nationalité dans le monde arabe des années 1830 aux années 1960: Négocier les appartenances et le droit," *Revue des mondes musulmans et de la Méditerranée* 137 (2015): 13–28.

47. A. Ṭawīlī, *al-Jinirāl Ḥusayn, Ḥayātu-hu wa āthāru-hu* (Tunis: Baladīyat Tunis, 1994), 101–6; Husayn Ibn 'Abdallah, *al-Qirṭāṣ al-mustaqīm fī ikhtilāl al-ḥukm bi-nafī jansiyya al-qā'id Nassīm* (Tunis: [n.pub.], 1878) and its translation into French, *Lettre du général Heusseïn au collège de la défense du gouvernement tunisien, dans l'affaire du caïd Nessim Samama* (Paris: Vve Renou, Maulde et Cock, 1878).

48. Husayn, *Al-Qirṭāṣ al-mustaqīm*, 236; and *Lettre du général Heusseïn*, 6.

49. In Arabic: "al-muslimūna rawu anna banī Adam haythu kānū min naw'-wāhid lā yaḥsan yajami'ūna ilā al-dīn al-wāḥid." Husayn, *Al-Qirṭāṣ al-mustaqīm*, 236; and *Lettre du général Heusseïn*, 6.

50. T. Eich, "The Forgotten Salafī-Abū l-Hudā As-Sayyādī," 61, 86; Mahmoud Haddad, "The Rise of Arab Nationalism Reconsidered," *International Journal of Middle East Studies* 26, no. 2 (1994): 201–22; Hasan Kayalı, *Arabs and Young Turks: Ottomanism, Arabism, and Islamism in the Ottoman Empire, 1908–1918* (Berkeley: University of California Press, 1997); Rashid Khalidi, "Ottomanism and Arabism in Syria before 1914: A Reassessment," in *The Origins of Arab Nationalism*, ed. Rashid Khalidi, Lisa Anderson, Muhammad Muslih, and Reeva S. Simon (New York: Columbia University Press, 1991), 58–69.

51. TNA, C. 11, d. 113, doc. 8765, Ottoman Consul to French Consul, July 2, 1887.

52. M'hamed Oualdi, "L'Héritage du général Husayn: La pertinence du 'national' et de la 'nationalité' au début du protectorat français sur la Tunisie," in *Penser le national au Maghreb et ailleurs*, ed. Fatma Ben Slimane and Hichem Abdessamad (Tunis: Arabesques, 2012), 65–88.

53. TNA, C. 11, d. 113, doc. 8775, French Consul to French Foreign Office, Florence, July 6, 1887.

54. TNA, C. 11, d. 113, doc. 8825–26, Khayr al-Din to *Bey*, September 4, 1887; BOA, HR.H . . . 171/13, Imperial Ottoman Embassy to Minister, September 24, 1887. Khayr al-Din intervened to avoid any new interference from the Ottoman consul in Florence because Husayn was a "manumitted slave of the Tunisian governor" and as such, his legacy belonged to his master.

55. TNA, C. 11, d. 113, doc. 8878, David Santillana to Secretary General Régnault, Florence, November 6, 1887; and doc. 8908, French Consul to French Foreign Office, Florence, November 12, 1887; Jean-Louis Bacqué-Grammont et Aksel Tibet, eds., *Cimetières et traditions funéraires dans le monde islamique* (Ankara: Türk Tarih Kurumu Basımevi, 1996), 215.

56. MAE (Nantes), 227 PO/1/100, Husayn's file (1887–88), Léon Elmilik to French Consul, July 8, 1887.

57. BOA HR.H . . . 171/13, Copy of Memorandum Received from the Office of the Grand Vezirate, November 5, 1887 (trans. Nilüfer Hatemi).

58. Ṭawīlī, *al-Jinirāl Ḥusayn*, 292; Bacqué-Grammont and Tibet, *Cimetières et traditions funéraires*, 215.

59. Mohammed el Aziz Ben Achour, " 'Tourbet el Bey' sépulture des Beys et de la famille husaynite à Tunis," *IBLA* 48, no. 155 (1985): 45–84.

60. TNA, C. 11, d. 113, doc. 8890, French Foreign Affairs to "résident général," Paris, November 14, 1887, f. 1.

61. BOA, HR.H . . . 171/13, Ottoman Consul in Livorno to Saïd Pacha, Minister of Ottoman Foreign Affairs, Livorno, July 2, 1887.

62. *L'Unione*, July 7, 1887.

63. Leor Halevi, "Funerary Practices," in Encyclopaedia of Islam, Three, ed. Kate Fleet et al., accessed February 12, 2019, http://dx.doi.org/10.1163/1573-3912_ei3_COM_27205.

64. TNA, C. 11, d. 113 doc. 8873, David Santillana, October 29, 1887; BOA, HR.H . . . 171/13, Copy of memorandum received from the Office of the Grand Vezirate, November 5, 1887 (trans. Nilüfer Hatemi).

65. TNA, C. 11, d. 113, doc. 885, Cable (no specific author), November 11, 1887.

66. BOA, HR.H . . . 171/13, Copy of memorandum received from the Office of the Grand Vezirate, November 5, 1887 (trans. Nilüfer Hatemi).

67. Achille Mbembé and Libby Meintjes, "Necropolitics," *Public Culture* 15, no. 1 (2003): 11. On the issue of "burial rites" and Tunisian nationalism, see Mary Dewhurst Lewis, "Over Our Dead Bodies. Burial Rites and Sovereignty in 1930s Tunisia," chap. 5 in *Divided Rule*, 145–78.

68. Lewis, *Divided Rule*, 132.

69. Al-Sanusi was jailed as a result of this action. Dupont, "An Expression of Pan-Islamism;" Leïla Blili, "Deux communautés face à la mort ou les griefs des a'yans contre la municipalité de Tunis," in *Les Relations tuniso-françaises au miroir des élites (XIX*e*-XX*e *siècles)*, ed. Noureddine Dougui (Tunis: Faculté des Lettres–La Manouba, 1997), 169–76.

70. Archives Nationales d'Outre-Mer (Aix-en-Provence), 26H13, 1882–89, Monthly Report on the Tunisian Situation, General Gillon to general governor of Algeria, March 1887.

71. Paul Sebag, *Tunis: Histoire d'une ville* (Paris: L'Harmattan, 1998), 131, 175, 235, 277; and *L'Évolution d'un ghetto nord-africain: La Ḥara de Tunis* (Paris: Presses universitaires de France, 1959), 79.

72. Nile Green, "Spacetime and the Muslim Journey West: Industrial Communications in the Making of the 'Muslim World,'" *American Historical Review* 118, no. 2 (2013): 423; Humayun Ansari, "'Burying the Dead:' Making Muslim Space in Britain," *Historical Research* 80, no. 210 (2007): 545–66.

73. Husayn, Letter 10, September 1868, in *Rasāʾil*, 1:79; M. ibn ʿUthmān Sanūsī, *al-Riḥla al-Ḥijāzīya* (Tunis: al-Sharikah al-Tūnisīyah lil-Tawzīʿ, 1976), 228.

74. Husayn, Letter 10, September 1868, in *Rasāʾil*, 1:79.

75. Muhammad al-Sanūsī, *al-Riḥla al-Ḥijāzīya*, 228.

76. MAE (Nantes), 227 PO/1/100, Husayn's file (1887–88), Léon Elmilik to Consul, Livorno, July 8, 1887.

77. TNA, C. 11, d. 113, doc. 8908, French Consul to French Foreign Office, Florence, November 12, 1887. On the Muslim cemetery of Livorno, see Mathieu Grenet, "Enterrer les 'Turcs' à Livourne en 1762: Sépultures étrangères, espaces littoraux et ségrégation spatiale," in *Les étrangers et les littoraux à l'époque moderne*, ed. François Brizay and Thierry Sauzeau (Rennes: Presses universitaires de Rennes, in press); Guillaume Calafat and Cesare Santus, "Les Avatars du 'Turc.' Esclaves et commerçants musulmans en Toscane (1600–1750)," in *Les Musulmans dans l'histoire de l'Europe*, ed. Jocelyne Dakhlia and Bernard Vincent (Paris: Albin Michel, 2011), 473; Paolo Castignoli, "La tolleranza: enunciazione e prassi di una regola di convivenza," in *Livorno, dagli archivi alla città: studi di storia*, ed. Lucia Frattarelli Fisher and Maria Lia Papi (Livorno: Belforte, 2001), 80; Gianfranco Grossi, "C'era una volt ail Cimitero de'Turchi," *Il Tirreno*, November 30, 2005.

78. MAE (Nantes), 227 PO/1/100bis, Husayn's legacy, June 30, 1888; Halevi, "Funerary Practices."

79. Halevi, "Funerary Practices"; Nicolas Vatin, "L'Inhumation intra-muros à Istanbul à l'époque ottoman," in *Les Ottomans et la mort: Permanences et mutation*, ed. Gilles Veinstein and Nicolas Vatin (Leiden: Brill, 1996), 166.

80. Muhammad Sanūsī, *al-Riḥla al-Ḥijāzīya*, 228; Halevi, "Funerary Practices"; Nicolas Vatin and Gilles Veinstein, "Les Obsèques des sultans ottomans," in *Les Ottomans et la mort*, 216, 218; Nadia Abu-Zahra, *The Pure and Powerful: Studies in Contemporary Muslim Society* (Reading, Berkshire: Ithaca Press, 1997), 48, 50, 58, 67–68.

81. TNA, C. 11, d. 113, doc. 8908, French Consul to French Foreign Office, Florence, November 12, 1887; and doc. 8890, French Foreign Affairs to "résident général," Paris, November 14, 1887, f. 1.

82. TNA, C. 11, d. 113, doc. 8908, French Consul to French Foreign Office, Florence, November 12, 1887.

83. BOA, HR.H . . . 171/13, Ottoman Consul to Saïd Pacha, Minister of Ottoman Foreign Affairs, Livorno, November 23, 1887: "le contenu de la caisse est figuré comme objet d'antiquité (afin d'éviter la moindre inquiétude à l'équipage)."

84. BOA, HR.H . . . 171/13, Cable from Yanko Photiades Pacha, Ottoman Ambassador Roma to Saïd Pacha, Minister of Ottoman Foreign Affairs, Rome, November 22, 1887; TNA, C. 11, d. 113, doc. 8907, French Consul to French Foreign Office, Florence, November 11, 1887; TNA, C. 11, d. 113, doc. 8890, French Foreign Affairs to "résident

général," Paris, November 14, 1887, f. 1; TNA, C. 11, d. 113, doc. 8908, French Consul to French Foreign Office, Florence, November 12, 1887; TNA, C. 11, d. 99, doc. 6, 'Amr Bu Hajib to Tunisian Prime Minister Muhammad Bu 'Attur, June 16, 1891.

85. Bacqué-Grammont and Tibet, *Cimetières et traditions funéraires*, 215: **1** feriqân-ı 'asâkir-I şâhaneden **2** ve ümerâ-yı Tûnusiyyeden olub **3** mû'ahhar^{en} Şurâ-yı Devlet **4** a'ẓ âlığina mê'mûr buyrulmuş **5** olan şâyeste-i maġfiret **6** ve sezâvâr-ı 'afv ü merḥamet **7** cenâb-ı Hüdâ Hüseyin Paşa **8** rûḥı-içün el-fatiḥa **9** sene 1305 ["Division General in the Imperial Army and commander at Tunis. Later appointed member of the Council of State deserving the pardon and meriting the forgiveness and compassion of the Lord God. In the year 1305."]

86. Emmanuel Fureix, *La France des larmes: Deuils politiques à l'âge romantique (1814–1840)* (Seyssel: Champ Vallon, 2009).

87. MAE (Nantes), 1TU/1V/1495, Cheykh Bayram's file, "Résident général" to Foreign Affairs, "Résidence générale," Tunisian Protectorate, December 22, 1889.

88. *La Dépêche Tunisienne*, March 7, 1890.

89. Ṭawīlī, *al-Jinirāl Ḥusayn*, 281.

90. Ṭawīlī, *al-Jinirāl Ḥusayn*, 284.

91. Ṭawīlī, *al-Jinirāl Ḥusayn*, 293.

92. Leon Carl Brown, *The Tunisia of Ahmad Bey, 1837–1855* (Princeton, NJ: Princeton University Press, 2016), 305–10; Odile Moreau, "Une guerre 'sur les marges.' La Tunisie et la guerre de Crimée par-delà l'oubli," in *Réformes de l'Etat et réformismes au Maghreb (XIXe–XXe siècles)*, ed. Odile Moreau (Paris: L'Harmattan-IRMC, 2009), 53–76; Andrew McGregor, "The Tunisian Army in the Crimean War: A Military Mystery," *Military History Online*, December 22, 2006, https://www.aberfoylesecurity.com/?p=1195.

93. Ṭawīlī, *al-Jinirāl Ḥusayn*, 294.

94. Ṭawīlī, *al-Jinirāl Ḥusayn*, 281, 285–86.

95. The scholarship on modern North Africa has underestimated these alternative forms of political action. See Catherine Mayeur-Jaouen, ed., *Saints et héros du moyen-orient contemporain* (Paris: Maisonneuve et Larose, 2002).

96. TNA, C. 11, d. 113, doc. 8786, David Santillana to "secrétaire général" Régnault, Florence, July 15, 1887; and doc. 9096, David Santillana to "secrétaire général" Régnault, Florence, January 28, 1888.

97. MAE (Nantes), 227 PO/1/1, French Consul to French Foreign Office, Florence, January 12, 1888.

98. TNA, C. 11, d. 113, doc. 8924, Chancellor Ghanem to Secretary General Régnault, Florence, December 22, 1887.

99. TNA, C. 11, d. 113, doc. 8930, Report on a meeting between the British, US, Greek, and Russian consuls, December 30, 1887.

100. TNA, C. 11, d. 113, doc. 8941, French Consultate to "résident général" Massicault, Florence, December 30, 1887; "L'incident de Florence," *Tunis-Journal*, January 4 and 7, 1888; *Unione*, January 10, 16, 19, 22, and 26, 1888.

101. Catherine Brice, *Histoire de l'Italie* (Paris: Perrin, 2003), 338–39. See also Lorenzo Del Piano, *La penetrazione italiana in Tunisi 1861–1881* (Padova: CEDAM, 1974), 101; Hugo Vermeren, *Les Italiens à Bône. Migrations méditerranéennes et colonisation de peuplement en Algérie (1865–1940)* (Paris: Nanterre University, 2015), 111, 107.

102. *L'Unione*, February 5, 1888; Çaycı, *La Question tunisienne*, 54–59.

103. TNA, C. 11, d. 113, doc. 9205, David Santillana to Secretary General Régnault, Florence. Santillana refers to *Revue de Droit International de Bruxelles* and *Journal du droit international privé*, 1888.

104. TNA, C. 11, d. 113, doc. 8925, French Consul to Prefect, Florence, December 22, 1887.

105. Daniela Melfa, *Migrando a Sud: coloni italiani in Tunisia, 1881-1939* (Rome: Aracne, 2008); Clancy-Smith, *Mediterraneans*.

106. Hanley, *Identifying with Nationality*, 175: "Egypt was a site of exception, a third place where they enjoyed special privileges."

5. Sovereigns, Mothers, and Creditors

1. Tunisian National Archives (TNA), C. 11, d. 113, doc. 8876, David Santillana to Secretary General, Florence, November 3, 1887.

2. Ministère des Affaires étrangères (MAE) (Nantes), 227 P)/1/100bis, Husayn's legacy, July 2, 1887: Note mentioning notary Carlo Querci.

3. Joseph Schacht, *Introduction au droit musulman* (Paris: Maisonneuve et Larose, 1999), 143; David S. Powers, *The Development of Islamic law and Society in the Maghrib: Qadis, Muftis, and Family Law* (Farnham, Surrey: Ashgate Variorum, 2011), 179; Pascale Ghazaleh, "Heirs and Debtors: Blood Relatives, Qur'anic Heirs, and Business Associates in Cairo, 1800–1850," in *Society and Economy in Egypt and the Eastern Mediterranean, 1600-1900: Essays in Honor of André Raymond*, ed. Nelly Hanna and Ra'ūf 'Abbās Ḥāmid (Cairo: American University in Cairo Press, 2005), 150.

4. MAE (Nantes), 227 PO/1/2 and 227 PO/1/100bis, Husayn's legacy, July 1887.

5. TNA, C. 11, d. 99, 'Amr Bu Hajib, June 18, 1891, f. 17; Husayn Ibn 'Abdallah, Letter 288, August 9, 1886, in *Rasā'il Ḥusayn ilā Khayr al-Dīn*, ed. Aḥmad 'Abd al-Sālam (Carthage: Bayt al-Ḥikma, 1992), 3:188.

6. Husayn, Letter 288, August 9, 1886, in *Rasāʾil*, 3:188.

7. MAE (Nantes), Tunisian Protectorate, 1TU/1V/1477, doc. 2, Prince Chadly to Résident Général Massicault, September 6, 1889.

8. Romain Bertrand, "Historicités indigènes et rapports vernaculaires au politique en 'situation coloniale,'" *Questions de recherche* 26 (2008): 26.

9. Amy Aisen Kallander, *Women, gender, and the palace households in Ottoman Tunisia* (Austin: University of Texas Press, 2013).

10. Daniel Schroeter, "Identity and Nation: Jewish Migrations and Intercommunity Relations in the Colonial Maghrib," in *La Bienvenue et l'adieu: Migrants juifs et musulmans au Maghreb, XVe-XXe siècles: Actes du colloque d'Essaouira migrations identité et modernité au Maghreb*, ed. Karima Dirèche and Frédéric Abécassis (Casablanca: Karthala, 2012), 1:125.

11. Jean Ganiage, *Les Origines du Protectorat français en Tunisie, 1861-1881*, 2nd ed. (Paris: Presses universitaires de France, 1969), 571.

12. *Journal Officiel tunisien*, December 22, 1887; TNA, C. 11, d. 113b, doc. 9380, Algerian Company to Secretary General of the Tunisian government, September 12,

1889; MAE (Nantes), Tunisian Protectorate, 1TU/1V/315, Résident Général Massicault's letters to individuals.

13. TNA, C. 11, d. 113, doc. 8757, To Eigenschenk, June 30, 1881.

14. TNA, C. 11, d. 97, doc. 84, David Santillana to Giaccomo Guttierez, Livorno, September 16, 1883; Ganiage, *Origines du Protectorat Français*, 602; Silvia Falconieri, "David Santillana," in *Dictionnaire des juristes ultramarins XVIIIe-XXe siècles*, ed. Florence Renucci (Paris: Mission de recherche Droit et justice, 2012), 272–73; F. Renucci, "David Santillana, acteur et penseur des droits musulman et européen," *Monde(s)* 1, no. 7 (2015): 25–44.

15. TNA, C. 11, d. 100, doc. 7523, "Summary of the affair, point about the claims," September 18, 1887; and d. 113, doc. 8768, Protectorate to French Foreign Affairs, July 4, 1887.

16. TNA, C. 11, d. 113b, doc. 9003, David Santillana to Secretary General Régnault, Florence, December 29, 1887.

17. TNA, C. 11, d. 113b, doc. 9003, David Santillana to Secretary General Régnault, Florence, December 29, 1887.

18. TNA, C. 11, d. 113b, doc. 9109, David Santillana to Résident Général Massicault, Florence, February 9, 1888, doc. 9109.

19. TNA, C. 11, d. 100, doc. 7523, "Summary of the affair, point about the claims," September 18, 1887.

20. TNA, C. 11, d. 113b, doc. 9145, Foreign Affairs, Political direction, Paris, April 7, 1888.

21. MAE (Nantes), Tunisian Protectorate, 1TU/1V/1071, Badoy to Résident Général Massicault, Tunis, August 5, 1893.

22. TNA, C. 11, d. 103, doc. 7728, Tunisian Prime Minister to Shaykh Maili Cadi, September 9, 1887; TNA, C. 11, d. 113, doc. 8823, Maliki Cadi to Prime Minister Bu 'Attur, September 11, 1887; TNA, C. 11, d. 113, doc. 8809, Maliki Cadi Muhammad al-Tahir al-Nayfar to Prime Minister, August 25, 1887; TNA, C. 11, d. 103, doc. 7729, Al-Hadi Zarruq to Prime Minister Muhammad al-'Aziz Bu 'Attur, Cairo, July 17, 1887; ibid., d. 98, doc. 90, Résident général to French Consul in Alexandria, May 1891.

23. TNA, C. 11, d. 103, doc. 7727, Tunisian Prime Minister to Al-Hadi Zarruq, September 7, 1887.

24. TNA, C. 11, d. 113, doc. 8819, French Consul to Résident Général, Florence, September 6, 1887; TNA, C. 11, d. 113, doc. 8821, French Consul to Résident Général, Florence, September 7, 1887; TNA, C. 11, d. 113, doc. 9017, Mustafa Bey to Léon Elmilik, January 8, 1888; TNA, C. 11, d. 113, doc. 35880–89, Agreement between Mustafa Bey and Léon Elmilik, August 17, 1887.

25. TNA, C. 11, d. 113, doc. 8911, Foreign Affairs to Résident Général Massicault, Paris, December 1, 1887.

26. TNA, C. 11, d. 113, doc. 8931, David Santillana to Résident Générale, Florence, December 26, 1887; and doc. 8933–34, *Beylical* decree (*amra*), December 25, 1887.

27. TNA, C. 11, d. 113, doc. 8969, Foreign Affairs to Résident Général Massicault, Paris, February 11, 1887; and doc. 8975, David Santillana to Secretary General, Florence, February 23, 1888.

28. TNA, C. 11, d. 113b, doc. 9109, David Santillana to Résident Général, Florence, February 9, 1888.

29. MAE (Nantes), Tunisian Protectorate, 1TU/1V/1217, "Note from the Central Police Station," March 28, 1894.

30. Antoine Perrier, "La liberté des protégés Souverains, ministres et serviteurs des monarchies marocaine et tunisienne sous protectorat français (1881–1956)," (Ph.D. diss., Sciences Po, 2019), 62–65; 'Adnān al-Manṣār, *Istrātijiyyā al-Haymana. Al-Ḥimāya al-Firansiyya wa Mu'assasāt al-Dawla al-Tūnisiyya*, 2nd ed. (Sfax: Maṭba'at al-tasfīr al-Fannī, 2005), 40, 51, 101–4.

31. TNA, C. 11, d. 100, doc. 7624, Agreement between David Santillana and Bertrand Meuville, Florence, August 12, 1888; and doc. 7541, Richard Nicolas to Résident Général Massicault, Tunis, November 26, 1888.

32. TNA, C. 11, d. 113b, doc. 9116, Bertrand Meuville to *Bey* of Tunis, New York, February 27, 1888.

33. FNA, Civil Sentence n°194, Eva Keusch, Angiolina Bertucci, De Laigue, and Léon Elmilik, Paoli, 1888.

34. Bu Hajib private archives (courtesy of Habib Bouhageb), Angiolina Bertucci's death certificate, n° 2241, 1914.

35. TNA, C. 11, d. 100, doc. 7584, David Santillana to Secretary General Régnault, Florence, August 22, 1888.

36. TNA, C. 11, d. 133, doc. 8799, 'Amr Bu Hajib to French Consul, Florence; and d. 113, doc. 8810, David Santillana to Secretary General Régnault, Florence, August 26, 1887.

37. TNA, C. 11, d. 100, doc. 7672, Eva Keusch to the Representative of the French Republic, Tunis, July 28, 1888.

38. TNA, C. 11, d. 100, doc. 7674, French Consul to Résident Général Massicault, Florence, December 29, 1887; and d. 113b, doc. 9141, David Santillana to Secretary General Régnault, Florence, April 20, 1888.

39. TNA, C. 11, d. 100, doc. 7674, French Consul to Résident Général Massicault, Florence, December 29, 1887; and d. 113b, doc. 9141, David Santillana to Secretary General Régnault, Florence, April 20, 1888.

40. TNA, C. 11, d. 100, doc. 7628, David Santillana to Secretary General Régnault, Florence, undated.

41. TNA, C. 11, d. 113b, doc. 7582, David Santillana to Secretary General Régnault, Florence, July 19, 1888.

42. TNA, C. 11, d. 100, doc. 7628, David Santillana to Secretary General Régnault, Florence, undated.

43. TNA, C. 11, d. 100, doc. 7582, David Santillana to Secretary General Régnault, Florence, July 19, 1888; and doc. 7528, David Santillana to Secretary General Régnault, Florence.

44. TNA, C. 11, d. 113b, doc. 7557, David Santillana to Secretary General Régnault, Florence, September 14, 1888.

45. TNA, C. 11, d. 113b, doc. 7567, David Santillana to Secretary General Régnault, Florence, September 8, 1888; and doc. 7552, David Santillana to Secretary General Régnault, September 14, 1888.

46. TNA, C. 11, d. 100, doc. 7566, Residence Générale to David Santillana, September 12, 1888.

47. TNA, C. 11, d. 100, doc. 7612, Agreement between Angiolina Bertucci and Tunisian Prime Minister Bu 'Attur, Tunis, September 11, 1889; d. 113b, doc. 9277, David Santillana to Secretary General Régnault, Florence, August 9, 1888; d. 100, doc. 7557, David Santillana to Secretary General Régnault, September 14, 1888.

48. TNA, C. 11, d. 113 b, doc. 7524, Direction Politique du Protectorat, Tunis, September 26, 1887; MAE (Courneuve), CPC, f. 628, Georges Benoit, Assistant Manager of the Protectorate to French Foreign Affairs, Tunis, September 26, 1887; Ali Noureddine, "Opposants et serviteurs: Les deux faces de la magistrature française dans la Régence de Tunis (1883–1886)," in *Contributions du séminaire sur les administrations coloniales (2009-2010)* (Paris: IHTP, 2011), 21.

49. TNA, C. 11, d. 100, doc. 7552, David Santillana to Secretary General Régnault, Florence, undated; doc. 7629, David Santillana to Secretary General Régnault, February 17, 1888; and doc. 7632, Richard Nicolas to Secretary General Régnault, Tunis, January 20, 1889.

50. TNA, C. 11, d. 100, doc. 7629, David Santillana to Secretary General Régnault, Florence, February 17, 1888.

51. TNA, C. 11, d. 100, doc. 7555, David Santillana to Secretary General Régnault, Florence, September 15, 1888.

52. TNA, C. 11, d. 100, doc. 7541, Richard Nicolas to Secretary General Régnault, Tunis, November 26, 1888.

53. TNA, C. 11, d. 100, doc. 7684, Death Certificate, February 27, 1887.

54. TNA, C. 11, d. 100, doc. 7629, David Santillana to Secretary General Régnault, Florence, February 17, 1888.

55. TNA, C. 11, d. 100, doc. 7612, Agreement between Angiolina Bertucci and Prime Minister Bu 'Attur, September 11, 1889.

56. *Al-Nuzha al-Khayriyya* 1306 (1888–89), 57.

57. TNA, C. 11, d. 100, doc. 7519, Translation of letter into Arabic, Angiolina Bertucci to David Santillana, September 8, 1888.

58. Jocelyne Dakhlia, "Entrées dérobées: L'historiographie du harem," *CLIO: Femmes, Genre, Histoire* 9 (1999), http://clio.revues.org/282; Leila Blili, *Histoire de familles. Mariages, répudiations et vie quotidienne à Tunis 1930–1975* (Tunis: Script, 1999), 56, 63–65 (about the mixed marriages); Blili, "Parenté et pouvoir dans la Tunisien houssaynite 1705–1957," unpublished manuscript, Tunis University, 2003–2004; Kallander, *Women, Gender, and the Palace Households*; Julia Clancy-Smith, "Women, Gender and Migration Along a Mediterranean Frontier: Pre-Colonial Tunisia, c.1815–1870," *Gender & History* 17, no. 1 (2005): 67.

59. See the case of Kmar Beya in Leila Blili, "Nazli Hanem, Kmar Bayya, and Khiriya Bin Ayyad: Three Women Living Between Istanbul, Cairo and Tunis in the Late Nineteenth and the Early Twentieth Centuries," in *Subversives and Mavericks in the Muslim Mediterranean: A Subaltern History*, ed. Odile Moreau and Stuart Schaar (Austin: University of Texas Press, 2013), 79–96. For the Egyptian case, see Kenneth Cuno, *Modernizing Marriage: Family, Ideology and Law in Nineteenth and Early Twentieth Century Egypt* (Syracuse, NY: Syracuse University Press, 2015), 43: the case of Khedive Abbas II, whose second wife was "Javidan (Djavidan) Hanim (1877–1968), the former Countess May Torok von Szendro, of Hungarian noble lineage."

60. Claude Liauzu, "Guerre des Sabines et tabou du métissage: Les mariages mixtes de l'Algérie coloniale à l'immigration en France," *Les cahiers du CEDREF* 8–9 (2000), http://journals.openedition.org/cedref/207, para. 33; Jocelyne Streiff-Fenart, "Le 'métissage' franco-algérien. Catégories politiques et imaginaires à propos des mariages mixtes," *Annuaire de l'Afrique du Nord* 29 (1990): 343.

61. Patricia M. E. Lorcin, "Colonialism and Imperialism: French North Africa," in *Encyclopedia of Women & Islamic Cultures* (Brill Online, 2016), http://dx.doi.org/10.1163/1872-5309_ewic_EWICCOM_0071d; Julia Clancy-Smith, "L'École Rue du Pacha, Tunis: L'enseignement de la femme arabe et 'la Plus Grande France' (1900–1914)," trans. Anne-Marie Engels-Brooks, *CLIO: Femmes, Genre, Histoire* 12 (2000), http://clio.revues.org/186; Ann Laura Stoler, "Tense and Tender Ties: The Politics of Comparison in North American History and (Post) Colonial Studies," *Journal of American History* 88, no. 3 (2001): 829–65; Elisa Camiscioli, "Women, Gender, Intimacy, and Empire," *Journal of Women's History* 25, no. 4 (2013): 141.

62. Christelle Taraud, "Les Femmes, le genre et les sexualités dans le Maghreb colonial (1830–1962)," *CLIO: Femmes, Genre, Histoire* 33 (2011): 160; Isabelle Surun, ed., *Les Sociétés coloniales à l'âge des Empires* (Paris: Atlande, 2012), 305.

63. Lorcin, "Colonialism and Imperialism"; Elizabeth Thompson, *Colonial Citizens: Republican Rights, Paternal Privilege, and Gender in French Syria and Lebanon* (New York: Columbia University Press, 2000).

64. TNA, C. 11, d. 113b, doc. 9133, Court of Accounts to Résident Général Justin Massicault, April 24, 1888; ʿAbd al-Qādir Vaḥḥah, *Al-Ḥaraka al-iṣlāḥīya al-zaytūnīya wa-al-madrasīya al-Tūnisīya: al-Shaykh Sālim Bū Ḥājib wa-manhajuhu al-iṣlāḥī, 1342-1244/1827–1828–1924* (Tunis: al-Ṭabʿah 1, 2007), 355; TNA, C. 11, d. 113b, doc. 8941, French consul to Résident Général Massicault, December 30, 1887.

65. TNA, C. 11, d. 99, ʿAmr Bu Hajib's report, June 18, 1891, f. 2: "fa-anā lā budda an atruku la-hum shay'an."

66. TNA, C. 11, d. 113, doc. 8941, French Consul to Résident Général Justin Massicault, Florence, December 30, 1887.

67. TNA, C. 11, d. 99, ʿAmr Bu Hajib's report, June 18, 1891; and d. 113b, doc. 8941, French Consul to Résident Général Justin Massicault, December 30, 1887.

68. TNA, C. 11, d. 113b, doc. 8941, French Consul to Résident Général Justin Massicault, December 30, 1887.

69. TNA, C. 11, d. 99, ʿAmr Bu Hajib, June 18, 1891, f. 17; and d. 113b, doc. 9133, Court of Accounts to Résident Général Justin Massicault, April 24, 1888.

70. TNA, C. 11, d. 98, doc. 52, French Vice-Consul in Halq al-Wadi to the French Résidence Representative about the baker Cardoso who owes 18,000 piasters for the *hegirian* years 1304–05 (1886–88), September 24, 1889; and d. 99, doc. 4, Minutes of the Clerk's Office, Civil Court, Tunis, March 28, 1889.

71. TNA, C. 11, d. 98, doc. 54, Massicault, Affair between Brulat and Husayn's Heirs, Vichy, August 11, 1889.

72. TNA, C. 11, d. 99, Amr Bu Hajib, June 18, 1891, f. 19.

73. TNA, C. 11, d. 99, Amr Bu Hajib, June 18, 1891, f. 19.

74. TNA, C. 11, d. 99, Amr Bu Hajib, June 18, 1891, f. 19.

75. TNA, C. 11, d. 99, Amr Bu Hajib, June 18, 1891, f. 19. The three *udūl* were Muhammad al-Shahid, al-ʿArussi Ibn ʿAyad (head of the Sadiqi College), and Mustafa

Lazughli (TNA, C. 11, d. 113, doc. 8941, French Consul in Florence to Résident Général Massicault, December 30, 1887).

76. TNA, C. 11, d. 113, doc. 8941, French Consul in Florence to Résident Général Massicault, December 30, 1887.

77. TNA, C. 11, d. 99, doc. 4, Minutes of the Clerk's Office, Civil Court, Tunis, March 28, 1889; and d. 113b, doc. 8958, Report on the Halq al-Wadi's buildings; Vaḥḥah, *al-Ḥaraka al-iṣlāḥīya al-zaytūnīya*, 356.

78. TNA, C. 11, d. 113b, doc. 8958, Report on the Halq al-Wadi buildings; and d. 99 'Amr Bu Hajib, June 18, 1891, f. 5.

79. TNA, C. 11, d. 99, doc. 4, Minutes of the Clerk's Office, Civil Court, Tunis, March 28, 1889; and d. 113b, doc. 8958, Report on the Halq al-Wadi buildings.

80. MAE (Nantes), Tunisian Protectorate, 1TU/1V/1477, doc. 2, Prince Chadly to Résident Général Massicault, September 6, 1889; TNA, C. 11, d. 99, doc. 8945, *mālikī bāsh muftī* (or main jurisconsult for the maliki school in Tunisia), Ahmad al-Sharif to 'Amr Bu Hajib, February 13, 1891.

81. TNA, C. 11, d. 99, 'Amr Bu Hajib's report, June 18, 1891, f. 19.

82. Madeline C. Zilfi, *The Politics of Piety: The Ottoman Ulema in the Postclassical Age (1600–1800)* (Minneapolis: Bibliotheca Islamica, 1988); Mahmoud Yazbak, "Nabulsi Ulama in the Late Ottoman Period, 1864–1914," *International Journal of Middle East Studies* 29, no. 1 (1997): 71–91; Arnold H. Green, *The Tunisian Ulama 1873–1915: Social Structure and Response to Ideological Currents* (Leiden: Brill, 1978).

83. Leon Carl Brown, *The Tunisia of Ahmad Bey, 1837–1855* (Princeton, NJ: Princeton University Press, 1974), 69–71, 75; James McDougall, *History and the Culture of Nationalism in Algeria* (Cambridge: Cambridge University Press, 2006); Green, *The Tunisian Ulama*; Fanny Colonna, *Les Versets de l'invincibilité: Permanence et changements religieux dans l'Algérie contemporaine* (Paris: Presses de la Fondation nationale des sciences politiques, 1995).

84. *al-Nuzha al-Khayriyya* 1306 (1888–89), 57.

85. TNA, C. 11, d. 113, doc. 8841, David Santillana to Secretary General Régnault, Florence, October 8, 1888.

86. TNA, C. 11, d. 113, doc. 8871, David Santillana to Secretary General Régnault, Florence, October 28, 1888; and d. 100, doc. 7523, "Summary of the affair, point about the claims," September 18, 1887.

87. TNA, C. 11, d. 113, doc. 8943, Résident général Massicault to David Santillana, January 1, 1888.

88. TNA, C. 11, d. 113, doc. 8888, David Santillana to Secretary General Régnault, Florence, November 13, 1887; and doc. 8921, French Consul to Prefect Gadda, Florence, December 16, 1887.

89. TNA, C. 11, d. 113b, doc. 9330, David Santillana to Secretary General Régnault, Florence, January 24, 1889, and *L'Unione*, August 5, 1888.

90. TNA, C. 11, d. 113, doc. 8756, Léon Elmilik to Prime Minister Bu 'Attur, Livorno, June 30, 1887.

91. TNA, C. 11, d. 113b, doc. 8936, Wire from David Santillana, undated.

92. TNA, C. 11, d. 113b, doc. 8938, Ghanem to Secretary General Régnault, Florence, undated; doc. 9351, French Consul Louis de Laigue to Résident Général Massicault, Florence, December 29, 1887; and doc. 8939, David Santillana to Résident Général Massicault, Florence, December 29, 1887.

93. TNA, C. 11, d. 113b, doc. 9078, French Foreign Office to Résident Général Massicault, Paris, January 22, 1888.

94. TNA, C. 11, d. 113b, doc. 9128, French Consul Louis de Laigue to French Foreign Affairs, Florence, March 12, 1888.

95. TNA, C. 11, d. 105, doc. 7825, Secretary General Régnault, December 24, 1890.

96. Schroeter, "Identity and Nation," 126

97. TNA, C. 11, d. 98, doc. 12, Agreement between 'Ali Bey and the Tunisian Government; TNA, C. 11, d. 113b, doc. 8967, Note on Angiolina Bertucci's defence, Civil Court, Florence; and d. 113, doc. 8900, French Consul to Foreign Affairs, Florence, November 22, 1887.

98. Schroeter, "Identity and Nation," 125.

99. Jessica Marglin, "Between Tolerance and Persecution: North Africans on North African Jewish History," in *After Orientalism: Critical Perspectives on Western Agency and Eastern Re-Appropriations*, ed. François Pouillon and Jean-Claude Vatin (Leiden: Brill, 2014), 64–65.

100. Ethan B. Katz, Lisa Moses Leff, and Maud S. Mandel, "Introduction: Engaging Colonial History and Jewish History," in *Colonialism and the Jews* (Bloomington: Indiana University Press, 2017), 2. See also Benjamin Stora, *Les Trois Exils: Juifs d'Algérie* (Paris: Stock, 2006).

101. TNA, C. 11, d. 113b, doc. 9154, Léon Elmilik to David Santillana, Florence, May 2, 1888.

102. Katz, Moses Leff, and Mandel, "Introduction," 9.

103. Joshua Schreier, *The Merchants of Oran: A Jewish Port at the Dawn of Empire* (Stanford, CA: Stanford University Press, 2017), 7.

104. Schreier, *The Merchants of Oran*, 7.

105. Marglin, "Between Tolerance and Persecution," 66, 68. On the same process, see Richard C. Parks, *Medical Imperialism in French North Africa: Regenerating the Jewish Community of Colonial Tunis* (Lincoln: University of Nebraska Press, 2017); Colette Zytnicki, "The 'Oriental Jews' of the Maghreb: Reinventing the North African Jewish Past in the Colonial Era," in *Colonialism and the Jews*, ed. Ethan B. Katz, Lisa Moses Leff, and Maud S. Mandel (Bloomington: Indiana University Press, 2017), 30.

106. Katz, Moses Leff, and Mandel, "Introduction," 9: "we still have much to learn about the processes by which Jews worked with, struggled against, supported, and benefited from the colonial order and the process of decolonization." Referring to Sarah Abrevaya Stein, "Protected Persons? The Baghdadi Jewish Diaspora, the British State, and the Persistence of Empire," *American Historical Review* 116, no. 1 (February 2011): 80–108.

107. Lucette Valensi, "Une Histoire des Juifs de Tunisie est-elle nécessaire? Est-elle posible?," in *Histoire communautaire, histoire plurielle: La communauté juive de Tunsie*, ed. Habib Kazdaghli (Tunis: CPU, 1999), 57, 61.

108. Taraud, "Les Femmes, le genre et les sexualités"; and James C. Scott, *La Domination et les arts de la résistance. Fragments du discours subalterne* (Paris: Editions Amsterdam, 2008), 11, 12, 199.

109. Francis Zimmermann, "Sahlins, Obeyesekere et la mort du capitaine Cook," *L'Homme* 146 (1998): 191–205; Romain Bertrand, "Politiques du moment colonial.

Historicités indigènes et rapports vernaculaires au politique en 'situation colonial,'" *Questions de recherche* 26 (2008): 10.

110. Angelo Torre, "'Faire communauté': Confréries et localité dans une vallée du Piémont (XVIIe–XVIIIe siècle)," *Annales. Histoire, Sciences Sociales* 62, no. 1 (2007).

111. Romain Bertrand, *Le long remords de la conquête: Manille-Mexico-Madrid, l'affaire Diego de Àvila, 1577–1580* (Paris: Éditions du Seuil, 2015), 311.

6. Husayn's Legacies in Colonial Tunisia

1. Lauren Benton, "Colonial Law and Cultural Difference: Jurisdictional Politics and the Formation of the Colonial State," *Comparative Studies in Society and History* 41, no. 3 (1999): 563–88; Mary Dewhurst Lewis, "Geographies of Power: The Tunisian Civic Order, Jurisdictional Politics, and Imperial Rivalry in the Mediterranean, 1881–1935," *Journal of Modern History* 80, no. 4 (2008): 791–830.

2. Ehud R. Toledano, "The Emergence of Ottoman-Local Elites (1700–1900): A Framework for Research," in *Middle Eastern Politics and Ideas: A History from Within*, ed. Ilan Pappé and Moshe Ma'oz (London: Tauris Academic Studies, 1997), 145–62.

3. Jane Hathaway, *The Politics of Households in Ottoman Egypt* (New York: Cambridge University Press, 1997), 17.

4. Leslie P. Peirce, *The Imperial Harem. Women and Sovereignty in the Ottoman Empire* (New York: Oxford University Press, 1993), 285.

5. Karim Rahem, *Le Sillage de la tribu: Imaginaires politiques et histoire en Algérie, 1843–1993* (Paris: Riveneuve, 2008).

6. Tunisian National Archives (TNA), C. 11, d. 113, doc. 8880, French Consulate, Expenses for Husayn's Inventory, October 27, 1887.

7. TNA, C. 11, d. 113, doc. 8820, French Consul to Foreign Affairs, Florence, September 7, 1887.

8. TNA, C. 11, d. 113, doc. 8898, Anonymous to French Consul, November 22, 1887, doc. 8898; doc. 8920, Chekri Ghanem to Secretary General Régnault, December 15, 1887; and doc. 9311, Chekri Ghanem to Secretary General Régnault, Florence, January 4, 1889.

9. TNA, C. 11, d. 101, doc. 7698, French Consul de Laigue to Résident Général Massicault, Florence, February 19, 1889.

10. TNA, C. 11, d. 113, doc. 9362, French Consul de Laigue to Foreign Affairs, Florence, February 20, 1889.

11. TNA, C. 11, d. 113, doc. 8820, French Consul de Laigue to Foreign Affairs, Florence, September 7, 1887; TNA, C. 11, d. 113, doc. 8853, Instructions to French Consul in Florence, Tunis, October 17, 1886; Ministère des Affaires étrangères (MAE) (Nantes), 227 PO/1/100bis, Husayn's legacy, October–November 1887; R. Imām (al-) Rashād. "Tārīkh al-Wazīr Muṣṭafā b. Ismā'īl," *al-Abḥāth* 17 (1969): 71–135.

12. MAE (Courneuve), CPC, file 629, Husayn's Legacy, Résident Général Massicault to French Foreign Office, Tunis, May 20, 1891; TNA, C. 11, d. 101, doc. 7704, Foreign Affairs to Résident Général Massicault, Paris, February 26, 1889.

13. Boubaker Sadok, *La Régence de Tunis au XVII*[e] *siècle: Ses relations commerciales avec les ports de l'Europe méditerranéenne, Marseille et Livourne* (Zaghouan: Ceroma, 1987); Francesca Trivellato, *The Familiarity of Strangers: The Sephardic Diaspora, Livorno, and Cross-Cultural Trade in the Early Modern Period* (New Haven, CT: Yale University Press, 2009); Linda Funaro, " 'Casa aperta, famiglia e traffico.' Presenze ebraiche e toscane fra Livorno e Tunisi," *Rassegna storica toscana* (2012): 93–117.

14. MAE (Courneuve), CPC, *bey* and *beylical* family, Political Direction of Protectorates to French Foreign Affairs Office, Paris, November 26, 1893; Mary Dewhurst Lewis, *Divided Rule: Sovereignty and Empire in French Tunisia, 1881-1938* (Berkeley: University of California Press, 2013), 28–29, 32, 56.

15. Sana Ben Achour, "Juges et magistrats tunisiens dans l'ordre colonial. 'Les juges musulmans' du Tribunal mixte immobilier de Tunisie (1886–1956)," in *La justice française et le droit pendant le protectorat en Tunisie*, ed. Nada Auzary-Schmaltz (Paris: Maisonneuve & Larose-Institut de recherche sur le Maghreb contemporain, 2007), 153–73; Arnold H. Green, *The Tunisian Ulama 1873-1915: Social Structure and Response to Ideological Currents* (Leiden: Brill, 1978), 145.

16. TNA, C. 11, d. 98, doc. 18, Giuseppe de Cardoso, trader from Genoa, Tunis, January 4, 1890.

17. Paolo Sartori, "Premessa. Waqf, colonialismo e pluralismo giuridico nelle società musulmane," *Quaderni storici* 132 (2009): 627–52.

18. Benton, "Colonial Law and Cultural Difference"; Lewis, "Geographies of Power."

19. A. Arazi and H. Ben-Shammay, "Mukhtaṣar," in *Encyclopaedia of Islam, Second Edition*, http://dx.doi.org/10.1163/1573-3912_islam_COM_0792.

20. Giorgio Levi Della Vida, "David Santillana," *Rivista degli studi orientali* 12 (1929–30): 453–61; Florence Renucci, "David Santillana, acteur et penseur des droits musulman et européen," *Monde(s)* 7, no. 1 (2015): 25–44; Silvia Falconiere, "David Santillana," in *Dictionnaire des juristes ultramarins XVIII*[e]*-XX*[e] *siècles*, ed. Florence Renucci (Paris: Mission de recherche Droit et justice, 2012), 272–76; David Santillana, *Istituzioni di diritto musulmano malichita con riguardo anche al sistema sciafiita*, 2 vols. (Rome: Istituto per l'Oriente, 1925–38).

21. TNA, C. 11, d. 113b, doc. 9019, David Santillana to Résident Général Massicault, no date; Inès Mrad Dali, "De l'esclavage à la servitude: Le cas des Noirs de Tunisie," *Cahiers d'études africaines* 179–80 (2005): 935–55; Ismael Musah Montana, *The Abolition of Slavery in Ottoman Tunisia* (Gainesville: University Press of Florida, 2013).

22. MAE (Nantes), Tunisian Protectorate, 1TU/1V/1071, War Ministry to Foreign Affairs, September 17, 1889; and Foreign Affairs to Résident Général Massicault, Paris, December 17, 1887.

23. TNA, C. 11, d. 103, doc. 7723, *Beylical* decree, April 6, 1890; Jacques Revault, *Palais et demeures de Tunis: XVIII*[e] *et XIX*[e] *siècles* (Paris: Editions du Centre national de la recherche scientifique, 1971), 229–32.

24. In the Algerian case, see Benjamin Claude Brower, *A Desert Named Peace: The Violence of France's Empire in the Algerian Sahara, 1844-1902* (New York: Columbia University Press, 2009); Yacine Daddi Addoun "L'Abolition de l'esclavage en Algérie: 1876–1871" (Ph.D. diss., York University, 2010).

25. Inès Mrad Dali, *Identités multiples et multitudes d'histoire: Les "Noirs tunisiens" de 1846 à aujourd'hui* (Paris: EHESS, 2009).

26. Pascale Ghazaleh, *Fortunes urbaines et stratégies sociales: Généalogies patrimoniales au Caire, 1780–1830* (Cairo: Institut français d'archéologie orientale, 2010).

27. Ghazaleh, *Fortunes urbaines et stratégies sociales*, 35, 44–46.

28. TNA, C. 11, d. 90, doc. 16, David Santillana to Secretary General Régnault, Florence, November 1, 1889.

29. TNA, C. 11, d. 98, doc. 12, Agreement between *Bey* and the Tunisian Government, including, on the first page, "une sentence d'un tribunal italien confirmée en appel et en cassation faisant défense à El Melik de réclamer quoi que ce soit à la succession Hussein avant d'avoir rendu ses comptes. En fait de comptes il n'existe qu'un seul registre où l'on constate que El Melik a touché plus qu'il ne lui revenait."

30. TNA, C. 11, d. 98, doc. 98, *Bey's* expenses for Husayn's estate, September 16, 1887; TNA, C. 11, d. 113b, doc. 9222, David Santillana to Secretary General Régnault, July 1888; MAE (Courneuve), CPC, file 629, Husayn's legacy, Résident Général Massicault to French Foreign Affairs Office, May 20?, 1891.

31. TNA, C. 11, d. 98, doc. 65, Eugène Rosa to Résident Général Massicault, Marseille, December, 22, 1890.

32. TNA, C. 11, d. 98, doc. 78, Paul Ricklin to Secretary General Régnault, Tunis, December 10, 1891; TNA, C. 11, d. 98, doc. 60, Moïse Elmilik to Résident Général Massicault, Tunis, December 16, 1891; MAE (Courneuve), CPC, file 629, Agreement between the *Bey* and the Tunisian Government; MAE (Courneuve), CPC, file 629, File about Léon Emilik's son, Agreement with David V. Elmilik.

33. TNA, C. 11, d. 98, doc. 47, Note on the Expenses for Husayn's Estate, October 12, 1890.

34. TNA, C. 11, d. 98, doc. 44, Receipts and Disbursements, March 10, 1892.

35. TNA, C. 11, d. 100, doc. 7521, Eva Keusch to Résident Général Massicault, Tunis, October 17, 1891. See also TNA, C. 11, d. 100, doc. 7522, Certificate from Dr. Bertholon, Tunis, December 2, 1891.

36. TNA, C. 11, d. 100, doc. 7522, Certificate from Dr. Bertholon, Tunis, December 2, 1891.

37. TNA, C. 11, d. 100, doc. 7643, Eva Keusch to Résident Genéral Massicault, Tunis, October 17, 1891.

38. TNA, C. 11, d. 100, doc. 7619, Résidence générale to David Santillana, July 12, 1888; TNA, C. 11, d. 100, doc. 7636, Agreement between the *Bey* and Eva Keusch, Florence, August 10, 1888; TNA, C. 11, d. 100, doc. 7606 Agreement on the Fr.50,000 note (*tadkīra*), Florence, September 8, 1888; MAE (Courneuve), CPC, file 629, Agreement between Léon Elmilik and Résident général Régnault, Tunis, March 28, 1890.

39. See chapter 2. MAE (Nantes), Tunisian Protectorate, 1TU/1V/1477, Husayn's legacy, doc. 2, List of *inzal*ists and what they owe, 1889s.

40. *La Dépêche tunisienne*, March 23, 1890.

41. ʿAbd al-Qādir Vaḥḥah, *Al-Ḥaraka al-iṣlāḥīya al-zaytūnīya wa-al-madrasīya al-Tūnisīya: al-Shaykh Sālim Bū Ḥājib wa-manhajuhu al-iṣlāḥī, 1342–1244/1827–1828–1924* (Tunis: al-Ṭabʿah 1, 2007), 108.

42. Green, *The Tunisian Ulama*, 130, 133; Moncef Chenoufi, "Les deux séjours de Muhammad ʿAbduh en Tunisie," *Les cahiers de Tunisie* 16 (1968): 71.

43. Majma' al-Tūnisī lil-'Ulūm wa-al-Ādāb wa-al-Funūn, *Al-Shaykh al-Muṣliḥ Sālim Bū Ḥājib wa-ishkālīyat al-'aṣr:* (Carthage: Bayt al-Ḥikmah, 2007), 19; Green, *The Tunisian Ulama*, 96.

44. Green, *The Tunisian Ulama*, 178–79.

45. J. Schacht, "Muḥammad 'Abduh," in *Encyclopaedia of Islam, Second Edition*, http://dx.doi.org/10.1163/1573-3912_islam_SIM_5378.

46. Green, *The Tunisian Ulama*, 148.

47. Green, *The Tunisian Ulama*, 147.

48. Chenoufi, "Les deux séjours de Muhammad 'Abduh," 76–78.

49. Aḥmad Ṭawīlī, *Al-Jinirāl Ḥusayn : ḥayātuhu wa-athāruhu* (Tunis: Baladīyat Tūnis, 1994), 28–29; MAE (Courneuve), CPC, file 630, Husayn's Legacy, March–April 1884, Envelope 5, Jamal al-Dīn to Husayn (three letters).

50. Green, *The Tunisian Ulama*, 167; Chenoufi, "Les deux séjours de Muhammad 'Abduh," 71.

51. Fanny Colonna, *Les Versets de l'invincibilité: Permanence et changements religieux dans l'Algérie contemporaine* (Paris: Presses de la Fondation nationale des sciences politiques, 1995); James McDougall, *History and the Culture of Nationalism in Algeria* (Cambridge: Cambridge University Press, 2006); Augustin Jomier, "Un Réformisme islamique dans l'Algérie coloniale: Oulémas ibadites et société du Mzab (c. 1880–c. 1970)" (Ph.D. diss., Le Mans University, 2015).

52. Chenoufi, "Les deux séjours de Muhammad 'Abduh en Tunisie," 65.

53. Mustapha Kraïem, "Au Sujet des incidences des deux séjours de Muhammad 'Abduh en Tunisie," *Revue d'histoire maghrébine* 3 (1975): 91–92.

54. Chenoufi, "Les deux séjours de Muhammad 'Abduh en Tunisie," 79.

55. Muhammed Maḥfūdh, *Tarājim al-mu'allifīn al-Tūnisiyyin* (Beirut: Dār al-Gharb al-Islāmī, 1982), 67; Al-Hādī Jallāb, *'Alī Bāsh Ḥānba, 1876–1918* (Tunis: Ministère de l'Enseignement Supérieur, 2005), 15.

56. Kenneth J. Perkins, *A History of Modern Tunisia* (New York: Cambridge University Press, 2004), 77.

57. Perkins, *A History of Modern Tunisia.*

58. Jallāb, *'Alī Bāsh Ḥānba*, 15; Andreas Tunger-Zanetti, *La Communication entre Tunis et Istanbul, 1860–1913: province et métropole* (Paris: L'Harmattan, 1996), 140–41; MAE (Nantes), 166 PO/C/280, 1830–1913, French Embassy in Istanbul; MAE (Nantes), 166 PO/C/280, 1881–1914, Tunisians in Turkey, file 84, 'Ali Bach Hamba, 1913–14.

59. Pierre Bardin, *Algériens et Tunisiens dans l'Empire Ottoman de 1848 à 1914* (Paris: CNRS, 1979), 160; M. Şükrü Hanioğlu, *A Brief History of the Late Ottoman Empire* (Princeton, NJ: Princeton University Press, 2008), 150.

60. Bardin, *Algériens et Tunisiens*, 166, 171; MAE (Nantes), Tunisians in Turkey, 1881–1914, 166 PO/C/280, file 84, Bach Hamba, 1913–1914, French embassy in Istanbul, 1830–1913.

61. Bardin, *Algériens et Tunisiens*, 160–61.

62. Taoufik Ayadi, *Mouvement réformiste et mouvements populaires à Tunis, 1906–1912* (Tunis: Publications de l'Université de Tunis, 1986), 65.

63. Jallāb, *'Alī Bāsh Ḥānba*, 15.

64. Tunger-Zanetti, *La Communication*, 141; Bardin, *Algériens et Tunisiens*, 230; MAE (Nantes), Tunisians in Turkey, 1881–1914, 166PO/C/280, file 84, Bach Hamba, 1913–1914, French embassy in Istanbul, 1830–1913.

65. Polat Safi, *The Ottoman Special Organization—Teşkilat-ı Mahsusa. An Inquiry into Its Operation and Administrative Characteristics* (Ankara: Bilkent University, 2012), 26, 215, 279.

66. Safi, *The Ottoman Special Organization*, 57–58, 61, 238.

67. Safi, *The Ottoman Special Organization*, 157; Polat Safi, "History in the Trench: The Ottoman Special Organization—*Teşkilat-ı Mahsusa* Literature," *Middle Eastern Studies* 48, no. 1 (2012): 93.

68. Safi, *The Ottoman Special Organization*, 10–11.

69. Jallāb, *'Alī Bāsh Ḥānba*, 50.

70. Jallāb, *'Alī Bāsh Ḥānba*, 15; Maḥfūdh, *Tarājim al-mu'allifīn al-Tūnisiyyin*, 67.

71. Tunger-Zanetti, *La Communication*, 141; MAE (Nantes), Tunisians in Turkey, 1881–1914, 166PO/C/280, file 84, Bach Hamba, 1913–1914, French embassy in Istanbul, 1830–1913; Bu Hajib private archives (courtesy of Habib Bouhageb), Angiolina Bertucci's death certificate, n° 2241, 1914.

72. Gilbert Meynier, *L'Algérie révélée: La guerre de 1914–1918 et le premier quart du XXe siècle* (Geneva: Droz, 1981); Benjamin Stora, *Messali Hadj, 1898–1974* (Paris: Sycomore, 1982); Hassine Raouf Hamza, *Communisme et nationalisme en Tunisie: De la "libération" à l'independance, 1943–1956* (Tunis: Université de Tunis, 1994).

73. Peter Heine, "Sālih ash-Sharīf at-Tūnisī, a North African Nationalist in Berlin During the First World War," *Revue de l'Occident musulman et de la Méditerranée* 33, no. 1 (1982): 92. See also Abdeljelil Temimi, "Min a'lāminā al-bārizīn wa al-mansiyyin, al-shaykh Ṣāliḥ al-Sharīf al- Tūnisī," *Revue d'histoire maghrébine* 23–24 (1981): 345–62.

74. Bardin, *Algériens et Tunisiens*, 195; Jonathan Derrick, *Africa's "Agitators": Militant Anti-Colonialism in Africa and the West, 1918–1939* (New York: Columbia University Press, 2008), 52.

75. Muhammad al-Fāḍil Ibn 'Ashūr, "Al-ṣadr Khalīl Bū Ḥājib," *al-Majalla al-Zaytūniyya* 9 (1955): 347–52.

76. Bu Hajib private archives (courtesy of Habib Bouhageb), *beylical* decree, December 31, 1904.

77. Bu Hajib private archives (courtesy of Habib Bouhageb), Annual report of the Tunisian Association, "La Musulmane."

78. *al-Nuzha al-Khayriyya* 1306 (1888–89), 57.

79. Ibn 'Ashūr, "Al-ṣadr Khalīl Bū Ḥājib," 351.

80. Vaḥḥah, *al-Ḥaraka al-iṣlāḥīya*, 386.

81. Barbara Degorge, "The Modernization of Education: A Case Study of Tunisia and Morocco," *European Legacy: Toward New Paradigms* 7, no. 5 (2002): 582.

82. Muḥammad al-Fāḍil Bin 'Ashūr, *Arkān al-nahḍa al-adabīya bi-Tūnis Maḥmūd Qābādū, Khayr al-Dīn, Sālim Bū Ḥājib* (Tunis: Maktabat al-Najāḥ, 1961), 50; Tunger-Zanetti, *La Communication*, 168; Abū al-Qāsim Muhammad Kirrū, *al-Amīra Nāzlī Fāḍila. Rā'iḍa al-Nahḍa fī Miṣr wa Tūnis* (Tunis: Dār al-Maghrib al-'Arabī, 2002), 133.

83. Ibn 'Ashūr, "Al-ṣadr Khalīl Bū Ḥājib," 350.

84. Tunger-Zanetti, *La Communication*, 168; Roger Allen, "Writings of Members of 'the Nāzlī Circle,'" *Journal of the American Research Center in Egypt* 8 (1969): 69; Bin 'Āshūr, *Arkān al-nahḍa al-Adabiya bi-Tūnis*, 50.

85. Tunger-Zanetti, *La Communication*, 117, 168; Allen, "Writings of Members of 'the Nāzlī Circle'"; Irène Melikoff, "Le Prince Mustafa Fazil et l'intelligentsia

ottomane," *Revue d'histoire maghrébine* 17 (1990): 121–24; Bin 'Āshūr, *Arkān al-nahḍa al-Adabiya bi-Tūnis*, 50.

86. Bin 'Ashūr, *Arkān al-nahḍa al-Adabiya bi-Tūnis*, 50; Tunger-Zanetti, *La Communication*, 68, 172.

87. Kirrū, *al-Amīra Nāzlī*, 150; Allen, "Writings of Members of 'the Nāzlī Circle,' " 79.

88. Mustapha Kraïem, "Autobiographie et relations avec la résidence de Tahar Khérédine Pacha," *Revue d'histoire maghrébine* 2 (1974): 185–91.

89. Frederick Cooper, "Conflict and Connection: Rethinking Colonial African History," *American Historical Review* 99, no. 5 (1994): 1533; Romain Bertrand, "Politiques du moment colonial. Historicités indigènes et rapports vernaculaires au politique en 'situation colonial,' " *Questions de recherche* 26 (2008): 27; Sherry B. Ortner, "Resistance and the Problem of Ethnographic Refusal," *Comparative Studies in Society and History* 37, no. 1 (1995): 173–93.

90. Beshara Doumani, ed., *Family History in the Middle East: Household, Property, and Gender* (Albany: State University of New York Press, 2003); Amy Aisen Kallander, *Women, Gender, and the Palace Households in Ottoman Tunisia* (Austin: University of Texas Press, 2013); Rahem, *Le Sillage de la tribu.*

91. Benton, "Colonial Law and Cultural Difference"; Lewis, "Geographies of Power."

92. Sartori, "Premessa."

Conclusion

1. James McDougall, "Crisis and Recovery Narratives in Maghrebi Histories of the Ottoman Period (ca. 1870–1970)," *Comparative Studies of South Asia, Africa and the Middle East* 31, no. 1 (2011): 131–48.

2. Lauren Benton, "Colonial Law and Cultural Difference: Jurisdictional Politics and the Formation of the Colonial State," *Comparative Studies in Society and History* 41, no. 3 (1999): 563–88; Mary Dewhurst Lewis, *Divided Rule: Sovereignty and Empire in French Tunisia, 1881-1938* (Berkeley: University of California Press, 2013); Paolo Sartori, "Premessa. Waqf, colonialismo e pluralismo giuridico nelle società musulmane," *Quaderni storici* 132 (2009): 627–52.

3. Isabelle Grangaud, "Prouver par l'écriture: Propriétaires algérois, conquérants français et historiens ottomanistes," *Genèses* 74, no. 1 (2009): 25–45; Simona Cerutti, "Histoire pragmatique, ou de la rencontre entre histoire sociale et histoire culturelle," *Tracés. Revue de Sciences humaines* 15 (2008): 147–68.

4. Leïla Blili, *Histoire de familles: Mariages, répudiations et vie quotidienne à Tunis, 1875-1930* (Tunis: Script, 1999); Karim Rahem, *Le Sillage de la tribu: Imaginaires politiques et histoire en Algérie, 1843-1993* (Paris: Riveneuve, 2008).

5. Fatma Ben Slimane, "Le Temps politique de la Tunisie 'moderne.' Héritage et questionnements," in *La Périodisation dans l'écriture de l'histoire du Maghreb*, ed. Fatma Ben Slimane and Hichem Abdessamad (Tunis: Arabesques, 2010): 49–63.

Bibliography

Archival Collections and Libraries Consulted

Aix-en-Provence

Archives Nationales d'Outre-Mer. Algeria 1H29, 26H13 (10), 26H13 (11).

Courneuve

Ministère français des Affaires étrangères (Courneuve). Correspondance politique et commerciale 1886–1917. Tunis files 1, 93, 460, 462, 602, 627–28, and 629–30 (Husayn's legacy).

Florence

Archivio di Stato: Repertorie sentenze corte d'appello 1886–89.
Archivio Notarile Provinciale: Registers 240–45.

Istanbul

Başbakanlık Osmanlı Arşivi: HR.H . . . 171/13.

Livorno

Archives of Livorno: Birth Certificate Register.
Biblioteca Labronica.

London

Public Record Office, Foreign Office 335/166; FO 102/170, 171, 172; FO 881/5594.

Lucca

Archivio di Stato.

Nantes

Ministère français des Affaires étrangères. Tunisia boxes 312–15, 901–2, 1071, 1179, 1180, 1211a, 1217, 1477, 1488, 1495; Florence boxes 1–2, 100, 100bis; Livorno boxes 131–33; Istanbul boxes 80, 84.

Rome

Ministero degli Affari Esteri, Archivio storico registers 1223–25, 1439–45.

Tunis

Tunisian National Archives. Historical Series box 11, files 97–114, Husayn's legacy.
Tunisian National Library. Newspapers: *Al-I'lām* (1887–88); *Le charivari tunisien* (1888–89); *Corriere Tunisino* (1887); *La dépêche tunisienne* (1889–91); *Le grelot tunisien* (1887–88); *Al-Ḥadīra* (1888); *Journal official tunisien* (1887–88); *Al-Nuzha al-Khayriyya* (1887); *Le Petit Tunisien* (1888–92); *Tunis-journal* (1887–88); *L'Unione* (1887–89).

Washington

National Archives and Records Administration. State Department's Central Files, Record Group 59, Microfilm T303, copies of correspondence between U.S. Consul in Tunis Amos Perry and the Secretary of State, William H. Seward, 1860–67; New York, Passenger and Crew Lists (including Castle Garden and Ellis Island), 1820–1957, Microfilm Serial: M237, 1820–97.

Printed Documents

al-Afghānī, Jamāl al-Dīn and 'Abduh, Muḥammad. *Al-'Urwa al-Wuthqā*, ed. Muḥammad Jamāl. Cairo: al-Maktaba al-ahliyya, 1927.
al-Imām, Rashād. "Tārīkh al-Wazīr Muṣṭafā b. Ismā'īl." *al-Abḥāth* 17 (1969): 71–135.
al-Tūnisī, Khayr al-Dīn. *Aqwām al-Masālik fī ma'rifa aḥwāl al-mamālik*. Carthage: Bayt al-Ḥikma, 2000.
——. *Essai sur les réformes nécessaires aux états musulmans*. Aix-en-Provence: Édisud, 1987.
Anonymous. "Le Bey de Tunis, l'Algérie, le Constitutionnel, la Vérité." *Bulletin de la société orientale* 4, no. 13–16 (1844): 82–96.
Ibn 'Abdallah, Husayn. *al-Qirṭāṣ al-mustaqīm fī ikhtilāl al-ḥukm bi-nafī jansiyya al-qā'id Nassīm*. Tunis: [n.pub.], 1878.
——. *Lettre du général Heusseïn au collège de la défense du gouvernement tunisien, dans l'affaire du caïd Nessim Samama*. Paris: Impr. de Vve Renou, Maulde et Cock, 1878.
——. *Rasā'il Ḥusayn ilā Khayr al-Dīn*. Ed. Ahmad 'Abd al-Sālam. 3 vols. Carthage: Bayt al-Ḥikma, 1991–92.
——. "Selection XII: Husayn Pacha. The Abolition of Slavery." In *Modern Arab Thought: Channels of the French Revolution to the Arab East*, ed. Khūrī Ra'īf, 152–57. Princeton, NJ: Kingston Press, 1983.
Jefferson, Thomas. *Notes on the State of Virginia*. Philadelphia: Prichard and Hall, 1788. In *Documenting the American South*. https://docsouth.unc.edu/southlit/jefferson/jefferson.html.
Santillana, David. *Istituzioni di diritto musulmano malichita con riguardo anche al sistema sciafiita*. 2 vols. Roma: Istituto per l'Oriente, 1926–38.
Sanūsī, Muḥammad ibn 'Uthmān. *Al-Riḥla al-Ḥijāzīya*. Tunis: al-Sharikah al-Tūnisīyah lil-Tawzī', 1976.

Secondary Sources

'Abd al-Mula, Mahmud. *Al-Madrasa al-Ḥarbiya bi-Bārdū*. Tunis: MTM, 2003.
'Abd al-Salām, Aḥmad. *Les Historiens tunisiens des XVIIe, XVIIIe et XIXe siècles: Essai d'histoire culturelle*. Paris: Librairie C. Klincksieck, 1973.
Abou El-Haj, Rifaat Ali. "The Ottoman Vezir and Paşa Households 1683–1703: A Preliminary Report." *Journal of the American Oriental Society* 94, no. 4 (1974): 438–47.
Abou-Hodeib, Toufoul. *A Taste for Home: The Modern Middle Class in Ottoman Beirut*. Stanford, CA: Stanford University Press, 2017.

Abu-Manneh, Butrus. "Sultan Abdulhamid II and Shaikh Abulhuda Al-Sayyadi." *Middle Eastern Studies* 15, no. 2 (1979): 131–53.
Abu-Zahra, Nadia. *The Pure and Powerful: Studies in Contemporary Muslim Society*. Reading, Berkshire: Ithaca Press, 1997.
Adda, Leila. "Les Apports culturels des réfugiés politiques italiens en Tunisie au XIXe siècle." In *Da maestrale e da scirocco: migrazioni attraverso il Mediterraneo atti del convegno, Facolta di scienze politiche, Catania, 23–25 gennaio 2003*, ed. Federico Cresti and Daniela Melfa, 65–82. Milano: A. Giuffré, 2006.
Ageron, Charles Robert. "Les Migrations des Musulmans algériencs et l'exode de Tlemcen (1830–1911)." *Annales* 22, no. 5 (1967): 1047–66.
Ago, Renata. *Gusto for Things. A History of Objects in Seventeenth-Century Rome*. Chicago: University of Chicago Press, 2013.
al-'Alāqī, 'Abd al-Karīm. "Khayr al-Dīn wa al-Yahūd fī Tūnis fī al-Qarn al-Tāsi' 'ashar." In *Khayr al-Dīn al-Tūnisī*, 149–81. Tunis: Maṭba'a al-Maghrib li-l-Nashr, 2011.
al-Manṣār, 'Adnān. *Istrātijiyyā al-Haymana. Al-Ḥimāya al-Firansiyya wa Mu'assasāt al-Dawla al-Tūnisiyya*. 2nd ed. Sfax: Maṭba'at al-tasfīr al-Fannī, 2005.
al-Maṭwī, Muḥammad al-Hādī. "Al-Jinirāl Ḥusayn yaktubu awwal qiṣṣa aṭfāl fī Tūnis." *al-Masār* 1 (1988): 114–20.
al-Sharīf, Muḥammad Farīd. *Sufarāʾ Tūnis al-Awāʾil*. Tunis: Imprimerie moderne, 2011.
Allen, Roger. "Writings of Members of 'the Nāzlī Circle.'" *Journal of the American Research Center in Egypt* 8 (1969): 79–84.
Alonso Acero, Beatriz. *Sultanes de Berbería en tierras de la cristiandad: exilio musulmán, conversión y asimilación en la monarquía hispánica, siglos XVI–XVII*. Barcelona: Edicions Bellaterra, 2006.
Amara, Noureddine. "Les Nationalités d'Amîna Hanım. Une pétition d'hérédité à la France (1896–1830)." *Revue des mondes musulmans et de la Méditerranée* 137 (2015): 49–72.
——. "La Nationalité des Touatis, un événement à la mesure d'empire (1830–1901)." *Maghreb et sciences sociales* (2012): 99–106.
Anastassiadou, Meropi. "Des Défunts hors du commun: Les possesseurs de livres dans les inventaires après décès musulmans de Salonique." *Turcica* 32 (2000): 197–252.
Anderson, Lisa. "Nineteenth-Century Reform in Ottoman Libya." *International Journal of Middle East Studies* 16, no. 3 (1984): 325–48.
Anheim, Étienne, and Olivier Poncet. "Fabrique des archives, fabrique de l'histoire." *Revue de Synthèse* 125, no. 1 (2004): 1–14.
Ansari, Humayun. "'Burying the Dead': Making Muslim Space in Britain." *Historical Research* 80, no. 210 (2007): 545–66.
Appadurai, Arjun. *The Social Life of Things: Commodities in Cultural Perspective*. Cambridge: Cambridge University Press, 1988.
Aprile, Sylvie. *Le Siècle des exilés: Bannis et proscrits de 1789 à la Commune*. Paris: CNRS, 2010.
Asad, Talal. "Conscripts of Western Civilization." In *Dialectical Anthropology: Essays in Honor of Stanley Diamond*, ed. Christine Ward Gailey, 333–51. Tallahassee: University Press of Florida, 1992.

Attal, Robert. *Le Caïd Nessim Samama de Tunis: Mécène du livre hébraïque*. Jérusalem: Attal, 1995.

—. *Kitve-'et ye-'itonim Yehudiyim bi-Tsefon Afrikah*. Tel-Aviv: Tel-Aviv University, 1996.

Ayadi, Taoufik. *Mouvement réformiste et mouvements populaires à Tunis, 1906–1912*. Tunis: Publications de l'Université de Tunis, 1986.

Aymes, Marc. *L'Art de l'état en Turquie: Arrangements de l'action publique de la fin de l'Empire ottoman à nos jours*. Paris: Karthala, 2013.

—. "Changeur d'empire." In *Penser, agir et vivre dans l'empire ottoman et en Turquie: Études réunies pour François Georgeon*, ed. Nathalie Clayer and Erdal Kaynar, 261–81. Paris-Louvain: Peeters, 2013.

—. "Defective Agency." In *Order and Compromise: Government Practices in Turkey from the Late Ottoman Empire to the Early Twenty-First Century*, ed. Marc Aymes, Benjamin Gourisse, and Elise Massicard, 25–44. Leiden: Brill, 2015.

—. *Un Grand progrès sur le papier. Histoire provinciale des réformes ottomanes à Chypre au XIX*e *siècle*. Louvain: Peeters, 2010.

Bacqué-Grammont, Jean-Louis, and Aksel Tibet, eds. *Cimetières et traditions funéraires dans le monde islamique*. Ankara: Türk Tarih Kurumu Basımevi, 1996.

Baldinetti, Anna. "David Santillana et Odoardo Maggiorani au service du Bey." *Revue d'histoire maghrébine* (1999): 71–75.

—. *David Santillana, l'uomo e il giurista (1855–1931): scritti inediti, 1878–1920*. Rome: Istituto per l'Oriente C.A. Nallino, 1995.

Bantigny, Ludivine. "Historicités du XXe siècle. Quelques jalons pour une notion." *Vingtième Siècle. Revue d'histoire* 117, no. 1 (2012): 13–25.

Bardin, Pierre. *Algériens et Tunisiens dans l'Empire Ottoman de 1848 à 1914*. Paris: CNRS, 1979.

Bargaoui, Sami. "Le Phénomène de l'usure dans la Tunisie de 1881." In *Réactions à l'occupation française de la Tunisie en 1881: Actes du 1er Séminaire d'histoire du Mouvement National: 29, 31 et 31 mai 1981, Hôtel Sidi Dhrif, Sidi Bou Saïd, Tunisie*, 353–57. Tunis: Ministère de l'enseignement supérieur et de la recherche scientifique, 1983.

—. " 'Quando dire è fare,' ovvero come ripetizione giuridica diventa una fonte storica." *Quaderni storici* 129, no. 3 (2008): 593–620.

—. "Les Titres fonciers dans la régence de Tunis à l'époque moderne: Interrogations autour d'une mutation documentaire." *IBLA* 208 (2011): 165–85.

Bargaoui, Sami, Isabelle Grangaud, and Simona Cerutti. Introduction to *Appartenance locale et propriété au nord et au sud de la Méditerranée*. Aix-en-Provence: Institut de recherches et d'études sur le monde arabe et musulman, 2015. https://books.openedition.org/iremam/3396?lang=en.

Bayart, Jean-François. "Les Études postcoloniales, une invention politique de la tradition?" *Sociétés politiques comparées. Revue européenne d'analyse des sociétés politiques* 14 (2009): 1–46.

Bayart, Jean-François, and Romain Bertrand. "De quel 'legs colonial' parle-t-on ?" *Esprit* 12 (2012): 134–60.

Becker, Howard Saul, and Charles C. Ragin, eds. *What Is a Case? Exploring the Foundations of Social Inquiry*. Cambridge: Cambridge University Press, 1992.

Bell, Dorian. *Globalizing Race: Antisemitism and Empire in French and European Culture*. Evanston, IL: Northwestern University Press, 2018.

Ben Achour, Mohammed el Aziz. " 'Tourbet el Bey' sépulture des Beys et de la famille husaynite à Tunis." *IBLA* 48, no. 155 (1985): 45–84.

Ben Achour, Sana. "Juges et magistrats tunisiens dans l'ordre colonial. 'Les juges musulmans' du Tribunal mixte immobilier de Tunisie (1886–1956)." In *La justice française et le droit pendant le protectorat en Tunisie*, ed. Nada Auzary-Schmaltz, 153–73. Paris: Maisonneuve & Larose-Institut de recherche sur le Maghreb contemporain, 2007.

Ben Slimane, Fatma. "Al-arḍ wa al-huwiyya, nushū' al-dawla al-turabiyya fī Tūnis 1881–1974." Tunis: Faculté des Sciences Humaines et Sociales de Tunis, 2009.

——. "Between Empire and Nation-State: The Problem of Borders in the Maghreb." In *Mediterranean Frontiers: Borders, Conflict and Memory in a Transnational Era*, ed. Dimitar Bechev and Kalypso Nicolaïdis, 35–55. London: Tauris Academic Studies, 2010.

——. "Une 'dhimma inversée?' La question des protections dans la Régence ottomane de Tunis." In *Les Musulmans dans l'histoire de l'Europe. II. Passages et contactes en Méditerranée*, ed. Jocelyne Dakhlia and Wolfgang Kaiser, 345–69. Paris: Albin Michel, 2013.

——. "Le Temps politique de la Tunisie 'moderne.' Héritage et questionnements." In *La Périodisation dans l'écriture de l'histoire du Maghreb*, ed. Fatma Ben Slimane and Hichem Adessamad, 49–63. Tunis: Arabesques, 2010.

Benbilghith, Chibani. "Le Rôle du général Hussein dans le mouvement réformiste en Tunisie pendant la deuxième moitié du XIXe siècle." *Arab Historical Review for Ottoman Studies* 11–12 (1995): 147–66 (Arabic), 224–27 (Abstract in French).

Benton, Lauren. "Colonial Law and Cultural Difference: Jurisdictional Politics and the Formation of the Colonial State." *Comparative Studies in Society and History* 41, no. 3 (1999): 563–88.

Bercher, Léon. "En marge du Pacte 'fondamental.' Un document inédit." *Revue tunisienne* (1939): 67–86.

Bernard, Augustin. "Le Recensement de 1906 en Algérie et en Tunisie." *Annales de Géographie*, 17, no. 91 (1908): 24–33.

Bertrand, Romain. *L'Histoire à parts égales: Récits d'une rencontre Orient-Occident, XVIe–XVIIe siècle*. Paris: Seuil, 2011.

——. "Historicités indigènes et rapports vernaculaires au politique en 'situation colonial.' " *Questions de recherche* 26 (2008): 2–49.

——. *Le long remords de la conquête: Manille-Mexico-Madrid, l'affaire Diego de Àvila, 1577-1580*. Paris: Éditions du Seuil, 2015.

——. "Au milieu du gué. Situation(s) de l'anthropologie historique du fait colonial." *Genèses* 97, no. 4 (2014): 140–48.

——. "Par-delà le grand récit de la Nation: L'identité nationale au prisme de l'histoire globale." *Savoir/Agir* 2 (2007): 51–60.

——. "Politiques du moment colonial. Historicités indigènes et rapports vernaculaires au politique en 'situation colonial.' " *Questions de recherche* 26 (2008): 1–49.

——. "Rencontres impériales. L'histoire connectée et les relations euro-asiatiques." *Revue d'histoire moderne et contemporaine* 54, no. 4bis (2007): 69–89.

Bessis, Juliette. *La Méditerranée fasciste. L'Italie mussolinienne et la Tunisie*. Paris: Karthala, 1981.

Bin 'Ashūr, Muḥammad al-Fāḍil. *Arkān al-nahḍa al-adabīya bi-Tūnis Maḥmūd Qābādū, Khayr al-Dīn, Sālim Bū Ḥājib*. Tunis: Maktabat al-Najāḥ, 1961.

Bin Sālim 'Umar. *Qābādū 1815-1871: ḥayātuhu wa-āthāruhu wa-tafkīruhu al-islāḥī*. Tūnis: Université de Tunis, 1975.

Blévis, Laure. "La Citoyenneté française au miroir de la colonisation: Étude des demandes de naturalisation des 'sujets français' en Algérie coloniale." *Genèses* 53, no. 4 (2003): 25–47.

Blili, Leila. "Deux communautés face à la mort ou les griefs des a'yans contre la municipalité de Tunis." In *Les Relations tuniso-françaises au miroir des élites (XIXe–XXe siècles)*, ed. Noureddine Dougui, 169–76. Tunis: Faculté des Lettres—La Manouba, 1997.

——. "Froufrous et bruissements: Costumes, tissus et couleurs dans la cour beylicale de Tunis au XIXe siècle." In *Trames de langue: Usages et métissages linguistiques dans l'histoire du Maghreb*, ed. Jocelyne Dakhlia, 223–39. Paris, Maisonneuve et Larose, 2004.

——. *Histoire de familles: Mariages, répudiations et vie quotidienne à Tunis, 1875-1930*. Tunis: Script, 1999.

——. "Nazli Hanem, Kmar Bayya, and Khiriya Bin Ayyad: Three Women Living Between Istanbul, Cairo and Tunis in the Late Nineteenth and the Early Twentieth Centuries." In *Subversives and Mavericks in the Muslim Mediterranean: A Subaltern History*, ed. Odile Moreau and Stuart Schaar, 79–96. Austin: University of Texas Press, 2013.

——. "Parenté et pouvoir dans la Tunisien houssaynite 1705–1957." Doctorat d'Etat, unpublished manuscript, last modified 2003–04.

Boccara, Elia. *In fuga dall'inquisizione: ebrei portoghesi a Tunisi. Due famiglie, quattro secoli di storia*. Firenze: Giuntina, 2011.

Boltanski, Luc, Elizabeth Claverie, Nicolas Offendstadt, and Stéphane Van Damme, eds. *Affaires, scandales et grandes causes: De Socrate à Pinochet*. Paris: Stock, 2007.

Borutta, Manuel, and Sakis Gekas. "A Colonial Sea: The Mediterranean, 1798–1956." *European Review of History: Revue européenne d'histoire* 19, no. 1 (2012): 1–13.

Boulu, Gilles. "Recherches sur les Scemama ou Samama de Tunis. Une dynastie de personnalités et de hauts-fonctionnaires de la régence de Tunis." Unpublished manuscript, https://docplayer.fr/17617165-Recherches-sur-les-scemama-ou-samama-de-tunis.html (March 2005).

Bouquet, Olivier. "Old Elites in a New Republic: The Reconversion of Ottoman Bureaucratic Families in Turkey (1909–1939)." *Comparative Studies of South Asia, Africa and the Middle East* 31, no. 3 (2011): 588–600.

——. *Les Pachas du sultan: Essai sur les agents supérieurs de l'État ottoman (1839-1909)*. Paris: Peeters, 2007.

——. "The Sultan's Sons-in-Law: Analysing Ottoman Imperial Damads." *Journal of the Economic and Social History of the Orient* 58, no. 3 (2015): 327–61.

Bouquet, Olivier, and Sinan Kuneralp. *Les Musurus, une famille de diplomates ottomans: Lettres et documents (1852-1920)*. Istanbul: Isis, 2015

Bousquet, Georges-Henri. *Les Berbères*. 4th ed. Paris: Presses universitaires de France, 1974.

Branche, Raphaëlle. *La Guerre d'Algérie: Une histoire apaisée?* Paris: Le Seuil, 2005.

——. " 'Au temps de la France.' Identités collectives et situation coloniale en Algérie." *Vingtième Siècle. Revue d'histoire* 117, no. 1 (2012): 199–213.

Bras, Jean-Philippe. "L'Imperfection de la propriété indigène, lieu commun de la doctrine juridique coloniale en Afrique du Nord." In *Appartenance locale et propriété au nord et au sud de la Méditerranée*. Aix-en-Provence: Institut de recherches et d'études sur le monde arabe et musulman, 2015. http://books.openedition.org/iremam/3441.

Brice, Catherine. "Confiscations et séquestres des biens des exilés politiques dans les États italiens au XIXe siècle. Questions sur une pratique et projets de recherches." *Diasporas. Circulations, migrations, histoire* 23–24 (2014): 147–63.

——. *Histoire de l'Italie*. Paris: Perrin, 2003.

Briganti, Lucilla. "L'emigrazione 'stagionale' dalla Toscana in Tunisia tra Ottocento e Novecento." In *Tunisia e Toscana*, ed. Vittorio Antonio Salvadorini, 154–57. Pise: Edistudio, 2002.

Brower, Benjamin Claude. *A Desert Named Peace: The Violence of France's Empire in the Algerian Sahara, 1844–1902*. New York: Columbia University Press, 2009.

Brown, Leon Carl. *The Tunisia of Ahmad Bey, 1837–1855*. Princeton, NJ: Princeton University Press, 1974.

Brown, Leon Carl, ed. *Imperial Legacy: The Ottoman Imprint on the Balkans and the Middle East*. New York: Columbia University Press, 1996.

Brunschvig, Robert. *Études d'islamologie*. Paris: Maisonneuve et Larose, 1976.

Burbank, Jane, and Frederick Cooper. *Empires in World History: Power and the Politics of Difference*. Princeton, NJ: Princeton University Press, 2010.

Burke III, Edmund. *Prelude to Protectorate in Morocco: Precolonial Protest and Resistance, 1860–1912*. Chicago: University of Chicago Press, 1976.

——. "Theorizing the Histories of Colonialism and Nationalism in the Arab Maghrib." *Arab Studies Quarterly* 20, no. 2 (1998): 5–19.

——. "Towards a History of the Maghrib." *Middle Eastern Studies* 11, no. 3 (1975): 306–23.

Calafat, Guillaume. "Une Mer jalousée. Juridictions maritimes, ports francs et régulation du commerce en Méditerranée (1590–1740)." Ph.D. diss., Paris 1 Panthéon-Sorbonne, 2013.

Calafat, Guillaume, and Cesare Santus. "Les Avatars du 'Turc.' Esclaves et commerçants musulmans en Toscane (1600–1750)." In *Les Musulmans dans l'histoire de l'Europe*, ed. Jocelyne Dakhlia and Bernard Vincent, 471–522. Paris: Albin Michel, 2011.

Camau, Michel. "Politique dans le passé, aujourd'hui au Maghreb." In *Les Trajectoires du politique. Volume II, la greffe de l'Etat*, ed. Jean-François Bayart, 63–93. Paris: Karthala, 1994.

Camiscioli, Elisa. "Women, Gender, Intimacy, and Empire." *Journal of Women's History* 25, no. 4 (2013): 138–48.

Cannon, Byron D. "Le marché de location des habous en Tunisie: Dialectique de développement agricole." In *Terroirs et Sociétés au Maghreb et Moyen Orient:*

Séminaire IRMAC, 1983-84, Table ronde franco-américaine CNRS/NSF, Lyon, juin 1984, ed. Byron D. Cannon, 79–108. Lyon: Maison de l'Orient, 1987.

Cannon, Byron D., and Muhammad Safiullah. "The Function of Wakalah in Intra-Ottoman Relations: Egypt and Tunisia in the First Half of the Nineteenth Century." *Arab Historical Review for Ottoman Studies* 5–6 (1992): 25–29.

Carlier, Omar. "L'Espace et le temps dans la recomposition du lien social: L'Algérie de 1830 à 1930." In *Urbanité arabe: Hommage à Bernard Lepetit*, ed. Jocelyne Dakhlia, 149–224. Paris: Sindbad, 1998.

Castignoli, Paolo. "La tolleranza: enunciazione e prassi di una regola di convivenza." In *Livorno: dagli archivi alla città: studi di storia*, ed. Lucia Frattarelli Fisher and Maria Lia Papi, 77–83. Livorno: Belforte, 2001.

Çayci, Abdurrahman. "La Question de la suppression des consulats tunisiens en Italie et l'entente italo-turque de 1863." *Cahiers de Tunisie* (1966): 41–52.

—. *La Question tunisienne et la politique ottomane, 1881-1913*. Ankara: Imprimerie de la Société turque d'histoire, 1992.

Çelik, Zeynep. *Empire, Architecture, and the City: French-Ottoman Encounters, 1830-1914*. Seattle: University of Washington Press, 2008.

Cerutti, Simona. *Étrangers: Étude d'une condition d'incertitude dans une société d'Ancien régime*. Paris: Fayard, 2012.

—. "Histoire pragmatique, ou de la rencontre entre histoire sociale et histoire culturelle." *Tracés. Revue de Sciences humaines* 15 (2008): 147–68.

Cerutti, Simona, and Anna-Malja Castrén. "Microhistory: Social Relations Versus Cultural Models?" In *Between Sociology and History: Essays on Microhistory, Collective Action, and Nation-Building*, ed. Anna-Maija Castrén, Markku Lonkila, and Matti Peltonen, 17–40. Helsinki: Finish Literature Society, 2004.

Çetin, Attilâ. "Un important mémoire de 1873 sur la Tunisie." *Studies on Ottoman Diplomatic History* (1987): 111–21.

—. "Relations ottomano-tunisiennes. Deux rapports sur les posititions française et italienne à propos de la mission de Khayreddin Pacha à Istanbul en 1871." *Studies on Turkish-Arab Relations* 4 (1989): 17–26.

Charpy, Manuel, and François Jarrige. "Introduction. Penser le quotidien des techniques. Pratiques sociales, ordres et désordres techniques au XIXe siècle." *Revue d'histoire du XIXe siècle* 45 (2012): 7–32.

Chater, Khalifa. *Insurrection et répression dans la Tunisie du XIXe siècle: La méhalla de Zarrouk au Sahel, 1864*. Tunis: Publications de l'Université de Tunis, 1978.

—. "Pensée en exil et décolonisation: Le cas maghrébin." *Cahiers de la Méditerranée* 82 (2011): 107–14.

Chatty, Dawn. "Refugees, Exiles, and Other Forced Migrants in the Late Ottoman Empire." *Refugee Survey Quarterly* 32, no. 2 (2013): 35–52.

Chenoufi, Ali. "Un Rapport inédit en langue arabe sur l'école de guerre du Bardo." *Cahiers de Tunisie* 95, no. 6 (1976): 45–118.

—. *Un Savant tunisien du XIXe siècle, Muḥammad as-Sanūsī: sa vie et son œuvre*. Tunis: Université de Tunis, 1977.

Chenoufi, Moncef. "Les deux séjours de Muhammad 'Abduh en Tunisie." *Les Cahiers de Tunisie* 16 (1968): 57–96.

Cherif, Mohamed-Hédi. *Pouvoir et société dans la Tunisie de H'usayn bin 'Ali, 1705–1740*. Tunis: Centre de publications universitaires, 2008.

Chih, Rachida, Catherine Mayeur-Jaouen, and Rüdiger Seesemann. "The Nineteenth Century: A Sufi Century?" In *Sufism, Literary Production, and Printing in the Nineteenth Century*, ed. Rachida Chih, Catherine Mayeur-Jaouen, and Rüdiger Seesemann, 3–24. Würzburg: Ergon Verlag, 2015.

Choate, Mark. "The Tunisia Paradox: Italy's Strategic Aims, French Imperial Rule, and Migration in the Mediterranean Basin." *California Italian Studies* 1 (2010): 1–20.

Christelow, Allan. *Algerians Without Borders: The Making of a Global Frontier Society*. Gainesville: University Press of Florida, 2012.

Clancy-Smith, Julia Ann. "Changing Perspectives on Colonialism and Imperialism: Women, Gender, Empire." In *Middle East Historiographies: Narrating the Twentieth Century*, ed. Israel Gershoni, Amy Singer, and Y. Hakan Erdem, 70–100. Seattle: University of Washington Press, 2006.

——. "L'École Rue du Pacha, Tunis: L'enseignement de la femme arabe et 'la Plus Grande France' (1900–1914)." Trans. Anne-Marie Engels-Brooks. *CLIO: Femmes, Genre, Histoire* 12 (2000). http://clio.revues.org/186.

——. *Mediterraneans: North Africa and Europe in an Age of Migration, c. 1800–1900*. Berkeley: University of California Press, 2011.

——. "Women, Gender and Migration Along a Mediterranean Frontier: Pre-Colonial Tunisia, c.1815–1870." *Gender & History* 17, no. 1 (2005): 62–92.

Clay, Christopher. "The Origins of Modern Banking in the Levant: The Branch Network of the Imperial Ottoman Bank, 1890–1914." *International Journal of Middle East Studies* 26, no. 4 (1994): 589–614.

Cleveland, William L. *The Making of an Arab Nationalist: Ottomanism and Arabism in the Life and Thought of Sati' al-Husri*. Princeton, NJ: Princeton University Press, 1971.

Clunet, Edouard. *Incident du consulat à Florence*. Paris: Marchal et Billard, 1888.

Coller, Ian. *Arab France: Islam and the Making of Modern Europe, 1798–1831*. Berkeley: University of California Press, 2011.

Colley, Linda. *The Ordeal of Elizabeth Marsh: A Woman in World History*. New York: Pantheon Books, 2007.

Colonna, Fanny. *Le Meunier, les moines et le bandit: Des vies quotidiennes dans l'Aurès, Algérie, du XXᵉ siècle*. Arles: Sindbad, 2009.

——. *Les Versets de l'invincibilité: Permanence et changements religieux dans l'Algérie contemporaine*. Paris: Presses de la Fondation nationale des sciences politiques, 1995.

Cooper, Frederick. "Conflict and Connection: Rethinking Colonial African History." *American Historical Review* 99, no. 5 (1994): 1516–45.

——. "What Is the Concept of Globalization Good For? An African Historian's Perspective." *African Affairs* 100 (2001): 189–213.

Cooper, Frederick, and Ann Laura Stoler. *Tensions of Empire: Colonial Cultures in a Bourgeois World*. Berkeley: University of California Press, 1997.

Coulson, Noel J. *Succession in the Muslim Family*. Cambridge: Cambridge University Press, 1971.

Cuno, Kenneth M. *Modernizing Marriage. Family, Ideology and Law in Nineteenth and Early Twentieth Century Egypt*. Syracuse, NY: Syracuse University Press, 2015.

——. *The Pasha's Peasants: Land, Society, and Economy in Lower Egypt, 1740–1858*. Cambridge: Cambridge University Press, 1992.
——. "Was the Land of Ottoman Syria Miri or Milk? An Examination of Juridical Differences within the Hanafi School." *Studia Islamica* 81 (1995): 121–52.
D'Agostini, Aldo. "De l'usage diplomatique du discours sur le panislamisme: La correspondance de l'ambassadeur français à Istanbul Charles-Joseph Tissot, lors de la crise tunisienne de 1881." *Archiv Orientalni* 81, no. 2 (2013): 149–72.
Daddi Addoun, Yacine. "L'Abolition de l'esclavage en Algérie: 1876–1871." Ph.D. diss., York University, 2010.
Dakhlia, Jocelyne, "1830, une rencontre?" In *Histoire de l'Algérie à la période coloniale (1830–1962)*, ed. Abderrahmane Bouchène and Sylvie Thénault, 142–49. Alger-Paris: Barzakh-La Découverte, 2012.
——. "Muslims in France: History Under the Carpet," in *Empire Lost: France and Its Other Worlds*, ed. Elisabeth Mudimbe-Boyi, 25–35. Lanham, MD: Lexington Books, 2009.
——. "Entrées dérobées: L'historiographie du harem." *CLIO: Femmes, Genre, Histoire* 9 (1999. http://clio.revues.org/282.
——. "Dans la mouvance du prince: La symbolique du pouvoir itinérant au Maghreb." *Annales* 43, no. 3 (1988): 735–60.
——. *L'Oubli de la cité: La mémoire collective à l'épreuve du lignage dans le Jérid tunisien*. Paris: Découverte, 1990.
——. *Tunisie, le pays sans bruit*. Arles: Actes sud, 2011.
Dakhlia, Jocelyne, and Bernard Vincent, eds. *Les Musulmans dans l'histoire de l'Europe*. Paris: Albin Michel, 2011.
Daly, M. W., and Carl F. Petry. *The Cambridge History of Egypt*. Cambridge: Cambridge University Press, 1998.
Davis, Natalie Zemon. *Trickster Travels: A Sixteenth-Century Muslim Between Worlds*. New York: Hill and Wang, 2006.
De Vito, Christian G. "Verso una microstoria translocale." *Quaderni storici* 3 (2015): 815–33.
Degorge, Barbara. "The Modernization of Education: A Case Study of Tunisia and Morocco." *European Legacy: Toward New Paradigms* 7, no. 5 (2002): 579–96.
Del Piano, Lorenzo. *La penetrazione italiana in Tunisia, 1861–1881*. Padova: A. Milani, 1964.
Deringil, Selim. "They Live in a State of Nomadism and Savagery: The Late Ottoman Empire and the Post-Colonial Debate." *Comparative Studies in Society and History* 45, no. 2 (2003): 311–42.
Derrick, Jonathan. *Africa's "Agitators." Militant Anti-Colonialism in Africa and the West 1918–1939*. New York: Columbia University Press, 2008.
Dodman, Thomas. "Un pays pour la colonie. Mourir de nostalgie en Algérie française, 1830–1880." *Annales* 66, no. 3 (2011): 743–84.
Doumani, Beshara. "Endowing Family: Waqf, Property Devolution, and Gender in Greater Syria, 1800 to 1860." *Comparative Studies in Society and History* 40, no. 1 (1998): 3–41.
Doumani, Beshara, ed. *Family History in the Middle East: Household, Property, and Gender*. Albany: State University of New York Press, 2003.

Dunn, Ross E. *Resistance in the Desert: Moroccan Responses to French Imperialism 1881–1912*. Madison: University of Wisconsin Press, 1977.

Dupont, Anne-Laure. "An Expression of Pan-Islamism in Tunisia at the Beginning of the French Protectorate: The Critique of the Tijāniyya in the Riḥla al-ḥijāziyya by Muḥammad b. ʿUthmān al-Sanūsī (d. 1900)." In *Sufism, Literary Production, and Printing in the Nineteenth Century*, ed. Rachida Chih, Catherine Mayeur-Jaouen, and Rüdiger Seesemann, 401–36. Würzburg: Ergon Verlag, 2015.

——. *Ǧurǧī Zaydān (1861–1914): Écrivain réformiste et témoin de la renaissance arabe*. Damas: IFPO, 2006.

——. "De la Demeure du Califat aux 'découvertes parisiennes': Muhammad al-Sanûsî (1851–1900), un lettré réformiste tunisien à l'épreuve du protectorat français." In *Penser, agir et vivre dans l'Empire ottoman et en Turquie. Etudes réunies pour François Georgeon*, ed. Nathalie Clayer and Erdal Kaynar, 47–65. Louvain: Peeters, 2013.

Eich, Thomas. "The Forgotten Salafī-Abū l-Hudā As-Sayyādī." *Die Welt des Islams* 43, no. 1 (2003): 61–87.

El Moudden, Abderrahmane, and ʿAbd al-Raḥīm Bin-Ḥāddah. *Al-ʿUthmāniyyūn fī al-Maghrib min khilāl al-arshīfāt al-maḥalliyya wa-al-mutawasiṭiyya*. Rabat: Al-Ṭabʿah 1, 2005.

Erdem, Y. Hakan. *Slavery in the Ottoman Empire and Its Demise, 1800–1909*. Basingstoke, Hampshire: Macmillan Press, 1996.

Ergene, Boğaā A. "Evidence in Ottoman Courts: Oral and Written Documentation in Early-Modern Courts of Islamic Law." *Journal of the American Oriental Society* 124, no. 3 (2004): 471–91.

Establet, Colette. *Familles et fortunes à Damas: 450 foyers damascains en 1700*. Damas: Institut français de Damas, 1994.

Establet, Colette, and Jean-Paul Pascual. "Les Inventaires après décès, sources froides d'un monde vivant." *Turcica* 32 (2000): 113–43.

Fahmy, Khaled. *In Quest of Justice: Islamic Law and Forensic Medicine in Modern Egypt*. Oakland: University of California Press, 2018.

Fakhfakh, Moncef. *Sommaire des registres administratifs et fiscaux aux archives nationales tunisiennes*. Tunis: Publication des Archives nationales tunisiennes, 1990.

Falconieri, Silvia. "David Santillana." In *Dictionnaire des juristes ultramarins XVIIIe–XXe siècles*, ed. Florence Renucci, 272–76. Paris: Mission de recherche Droit et justice, 2012.

Filippini, Jean-Pierre. "Les Juifs d'Afrique du Nord et la communauté de Livourne au XVIIIe siècle." In *Les Relations intercommunautaires juives en Méditerranée occidentale XIIIe–XXe siècle*, ed. Jean-Louis Miège, 50–63. Paris: CNRS, 1984.

Findley, Carter Vaughn. *Ottoman Civil Officialdom: A Social History*. Princeton, NJ: Princeton University Press, 1989.

Frierson, Elizabeth B. "Cheap and Easy: The Creation of Consumer Culture in Late Ottoman Society." In *Consumption Studies and the History of the Ottoman Empire, 1550–1922: An Introduction*, ed. Donald Quataert, 243–60. Albany: State University of New York Press, 2000.

Funaro, Linda. "'Casa aperta, famiglia e traffico.' Presenze ebraiche e toscane fra Livorno e Tunisi." *Rassegna storica toscana* (2012): 93–117.

——. "A Mediterranean Diaspora: Jews from Leghorn in the Second Half of the Nineteenth Century." In *L'Europe méditerranéenne/Mediterranean Europe*, ed. Marta Petricioli, 95–110. Bruxelles: Peter Lang, 2008.

Fureix, Emmanuel. *La France des larmes: Deuils politiques à l'âge romantique (1814–1840)*. Seyssel: Champ Vallon, 2009.

Gagné, Natacha. "The Study of Colonial Situations: The Emergence of a New General Approach?" *Reviews in Anthropology* 41, no. 2 (2012): 109–35.

Ganiage, Jean. *Les Origines du Protectorat français en Tunisie, 1861–1881*. Paris: Presses universitaires de France, 1959.

——. *La Population européenne de Tunis au milieu du XIXe siècle: Étude démographique*. Paris: Presses universitaires de France, 1960.

Garcia-Arenal, Mercedes. *A Man of Three Worlds: Samuel Pallache, a Moroccan Jew in Catholic and Protestant Europe*. Baltimore, MD: Johns Hopkins University Press, 2003.

Gardey, Delphine. *Écrire, calculer, classer: Comment une révolution de papier a transformé les sociétés contemporaines, 1800–1940*. Paris: La Découverte, 2008.

Gasperoni, Michael. " 'Une sorte de domaine naturel créé par la nature des choses.' L'émergence et la formalisation du jus gazagà dans les ghettos d'Italie à l'époque moderne." Unpublished manuscript, https://hal.archives-ouvertes.fr/halshs-01197089, last updated 2015.

Ghazaleh, Pascale. *Fortunes urbaines et stratégies sociales: Généalogies patrimoniales au Caire, 1780–1830*. Cairo: Institut français d'archéologie orientale, 2010.

——. "Heirs and Debtors: Blood Relatives, Qur'anic Heirs, and Business Associates in Cairo, 1800–1850." In *Society and Economy in Egypt and the Eastern Mediterranean, 1600–1900: Essays in Honor of André Raymond*, ed. Nelly Hanna and Ra'ūf 'Abbās Ḥāmid, 143–58. Cairo: American University in Cairo Press, 2005.

——. *Held in Trust: Waqf in the Islamic world*. Cairo: American University in Cairo Press, 2011.

Ghobrial, John Paul A. "The Secret Life of Elias of Babylon and the Uses of Global Microhistory." *Past & Present* 222 (2014): 51–93.

Gilsenan, Michael. "Possessed of Documents: Hybrid Laws and Translated Texts in the Hadhrami Diaspora." In *Ethnographies of Islam: Ritual Performances and Everyday Practices*, ed. Baudouin Dupret, Thomas Pierret, Paulo G. Pinto, and Kathryn Spellman-Poots, 181–92. Edinburgh: Edinburgh University Press, 2012.

——. "Topics and Queries for a History of Arab Families and Inheritance in Southeast Asia: Some Preliminary Thoughts." In *Southeast Asia and the Middle East: Islam, Movement, and the Longue Durée*, ed. Eric Tagliacozzo, 199–234. Stanford, CA: Stanford University Press, 2009.

——. "Translating Colonial Fortunes: Dilemmas of Inheritance in Muslim and English Laws across a Nineteenth-Century Diaspora." *Comparative Studies of South Asia, Africa and the Middle East* 31, no. 2 (2011): 355–71.

Giudice, Christophe. "La Construction de Tunis "ville européenne" et ses acteurs de 1860 à 1945." *Correspondances* 70 (2002): 9–16.

Grangaud, Isabelle. "Affrontarsi in archivo. Tra storia ottomana e storia coloniale (Algeri 1830)." *Quaderni storici* 129, no. 3 (2008): 621–52.

——. "Prouver par l'écriture: Propriétaires algérois, conquérants français et historiens ottomanistes." *Genèses* 74, no. 1 (2009): 25–45.

——. *La Ville imprenable: Une histoire sociale de Constantine au 18ᵉ siècle*. Paris: Éd. de l'École des hautes études en sciences sociales, 2002.
Grange, Daniel J. *L'Italie et la Méditerranée (1896–1911): Les fondements d'une politique étrangère*. Rome: Ecole française de Rome, 1994.
Green, Arnold H. *The Tunisian Ulama 1873–1915: Social Structure and Response to Ideological Currents*. Leiden: Brill, 1978.
Green, Nile. "Spacetime and the Muslim Journey West: Industrial Communications in the Making of the 'Muslim World.'" *American Historical Review* 118, no. 2 (2013): 401–29.
Grenet, Mathieu. "Enterrer les 'Turcs' à Livourne en 1762: Sépultures étrangères, espaces littoraux et ségrégation spatiale." In *Les étrangers et les littoraux à l'époque moderne*, ed. François Brizay et Thierry Sauzeau. Rennes: Presses universitaires de Rennes, in press.
Grossi, Gianfranco. "C'era una volt ail Cimitero de'Turchi." *Il Tirreno*, November 30, 2005.
Guignard, Didier. *L'Abus de pouvoir dans l'Algérie coloniale, 1880–1914: Visibilité et singularité*. Paris: Presses universitaires de Paris Ouest, 2010.
Haddad, Mahmoud. "The Rise of Arab Nationalism Reconsidered." *International Journal of Middle East Studies* 26, no. 2 (1994): 201–22.
Hanioğlu, M. Şükrü. *A Brief History of the Late Ottoman Empire*. Princeton, NJ: Princeton University Press, 2008.
Hanley, Will. *Identifying with Nationality. Europeans, Ottomans, and Egyptians in Alexandria*. New York: Columbia University Press, 2017.
Harber, Charles C. "Tunisian Land Tenure in the Early French Protectorate." *Muslim World* 63, no. 4 (1973): 307–15.
Harouvi, Yavel. "Les Conflits autour du testament du Caïd Nessim Scemama d'après quelques sources hébraïques." In *Entre Orient et Occident: Juifs et Musulmans en Tunisie*, ed. Denis Cohen-Tannoudji, 143–56. Paris: Éd. de l'Éclat, 2007.
Hathaway, Jane. *The Politics of Households in Ottoman Egypt: The Rise of the Qazdağlis*. New York: Cambridge University Press, 1997.
Hautcoeur, Pierre-Cyrille. "Les Transformations du crédit en France au XIXᵉ siècle." *Romantisme* 151, no. 1 (2011): 23–38.
Heine, Peter. "Sālih ash-Sharīf at-Tūnisī, a North African Nationalist in Berlin During the First World War." *Revue de l'Occident musulman et de la Méditerranée* 33, no. 1 (1982): 89–95.
Hénia, Abdelhamid. *Le Grid: Ses rapports avec le beylik de Tunis (1676–1840)*. Tunis: Publications de l'Université de Tunis, 1980.
——. *Propriété et stratégies sociales à Tunis, XVIᵉ–XIXᵉ siècles*. Tunis: Université Tunis I, 1999.
Hugon, Henri. "Le domaine de l'État en Tunisie." In *Conférences sur les administrations tunisiennes*, 239–77. Sousse: Imprimerie française, 1898.
Hurewitz, J. C. *The Middle East and North Africa in World Politics: A Documentary Record*. 2 vols. New Haven, CT: Yale University Press, 1975.
Ibn 'Ashūr, Muhammad al-Fāḍil. "Al-ṣadr Khalīl Bū Ḥājib." *al-Majalla al-Zaytūniyya* 9 (1955): 347–52.
——. *Arkān al-nahḍa al-adabīya bi-Tūnis Maḥmūd Qābādū, Khayr al-Dīn, Sālim Bū Ḥājib*. Tunis: Maktabat al-Najāḥ, 1961.

Inalcık, Halil, and Donald Quataert, eds. *An Economic and Social History of the Ottoman Empire, 1300–1914*. Cambridge: Cambridge University Press, 1994.

Isabella, Maurizio, and Konstantina Zanou, eds. *Mediterranean Diasporas: Politics and Ideas in the Long Nineteenth Century*. London: Bloomsbury Academic, 2016.

Islamoğlu, Huri. "Modernities Compared: State Transformations and Constitutions of Property in the Qing and Ottoman Empires." *Journal of Early Modern History* 5, no. 4 (2001): 353–86.

Jallāb, Al-Hādī. *'Alī Bāsh Ḥānba, 1876–1918*. Tunis: Ministère de l'Enseignement Supérieur, 2005.

Jamoussi, Habib. *Juifs et chrétiens en Tunisie au 19e siècle. Essai d'une étude socio-culturelle des communautés non-musulmanes (1815–1881)*. Sfax: Amal Éditions, 2010.

Jdey, Ahmed. *Ahmed ibn Abi Dhiaf: Son œuvre et sa pensée. Essai d'histoire culturelle*. Zaghouan: Fondation Temimi pour la recherche scientifique et l'information, 1996.

Jirousek, Charlotte. "The Transition to Mass Fashion System Dress in the Later Ottoman Empire." In *Consumption Studies and the History of the Ottoman Empire, 1550–1922: An Introduction*, ed. Donald Quataert, 201–41. Albany: State University of New York Press, 2000.

Johansen, Baber. "Formes de langage et fonctions publiques: Stéréotypes, témoins et offices dans la preuve par l'écrit en droit musulman." *Arabica* 44, no. 3 (1997): 333–76.

Jomier, Augustin. "Un Réformisme islamique dans l'Algérie coloniale: Oulémas ibadites et société du Mzab (c. 1880–c. 1970)." Ph.D. diss., Le Mans University, 2015.

Julien, Charles-André. "Colons français et Jeunes-Tunisiens, 1882–1912." *Revue française d'histoire d'outre-mer* 54, no. 194 (1967): 87–150.

Kallander, Amy Aisen. *Women, Gender, and the Palace Households in Ottoman Tunisia*. Austin: University of Texas Press, 2013.

Karoui, Hachmi. "La Résistance populaire à l'occupation française (1881) chez les élites tunisiennes." In *Connaissances du Maghreb: Sciences sociales et colonisation*, ed. Jean-Claude Vatin, 401–21. Paris: CNRS, 1984.

Karoui, Hachmi, and Ali Mahjoubi. *Quand le soleil s'est levé à l'Ouest: Tunisie 1881, impérialisme et résistance*. Tunis: Cérès Productions, 1983.

Karpat, Kemal H. "The Land Regime, Social Structure, and Modernization in the Ottoman Empire." In *Studies on Ottoman Social and Political History: Selected Articles and Essays*, 327–51. Leiden: Brill, 2002.

Katz, Ethan B., Lisa Moses Leff, and Maud S. Mandel. "Introduction: Engaging Colonial History and Jewish History." In *Colonialism and the Jews*, ed. Ethan B. Katz, Lisa Moses Leff, and Maud S. Mandel, 1–26. Bloomington: Indiana University Press, 2017.

Kayalı, Hasan. *Arabs and Young Turks Ottomanism, Arabism, and Islamism in the Ottoman Empire, 1908–1918*. Berkeley: University of California Press, 1997.

Kazdaghli, Habib. "The Case of Tunisia." In *A History of Jewish-Muslim Relations: From the Origins to the Present Day*, ed. Abdelwahab Meddeb and Benjamin Stora, trans. Jane Marie Todd and Michael B. Smith, 316–19. Princeton, NJ: Princeton University Press, 2013.

Kenbib, Mohammed. *Les Protégés: Contribution à l'histoire contemporaine du Maroc*. Casablanca: Faculté des Lettres et des Sciences Humaines, 1996.

Kertzer, David I., and Richard P. Saller. *The Family in Italy: From Antiquity to the Present.* New Haven, CT: Yale University Press, 1991.
Khalidi, Rashid. "Ottomanism and Arabism in Syria Before 1914: A Reassessment." In *The Origins of Arab Nationalism*, ed. Rashid Khalidi, Lisa Anderson, Muhammad Muslih, and Reeva S. Simon, 58–69. New York: Columbia University Press, 1991.
Khuri-Makdisi, Ilham. *The Eastern Mediterranean and the Making of Global Radicalism, 1860–1914.* Berkeley: University of California Press, 2010.
Kirrū, Abū al-Qāsim Muhammad. *al-Amīra Nāzlī Fāḍīla. Rā'iḍa al-Nahḍa fī Miṣr wa Tūnis.* Tunis: Dār al-Maghrib al-'Arabī, 2002.
Kraïem, Mustapha. "Autobiographie et relations avec la résidence de Tahar Khereddine Pacha." *Revue d'histoire maghrébine* 2 (1974): 185–91.
——. "Au Sujet des incidences des deux séjours de Muhammad 'Abduh en Tunisie." *Revue d'histoire maghrébine* 3 (1975): 91–94.
Kuneralp, Sinan. "Diplomates et consuls ottomans en France au XIXe siècle." In *L'Empire Ottoman, la République de Turquie et la France*, ed. Hâmit Batu and Jean-Louis Bacqué-Grammont, 305–13. Istanbul: ISIS, 1996.
Kunt, İ. Metin. *The Sultan's Servants. The Transformation of Ottoman Provincial Government, 1550–1650.* New York: Columbia University Press, 1983.
Labib, Malak. "Crise de la dette publique et missions financières européennes en Égypte, 1878–1879." *Monde(s)* 4 (2013): 23–43.
Lafi, Nora. *Une Ville du Maghreb entre ancien régime et réformes ottomanes: Genèse des institutions municipales à Tripoli de Barbarie, 1795–1911.* Paris: L'Harmattan, 2002.
Lambert, David. *Notables des colonies: Une élite de circonstance en Tunisie et au Maroc, 1881–1939.* Rennes: Presses universitaires de Rennes, 2009.
Landes, David S. *Bankers and Pashas: International Finance and Economic Imperialism in Egypt.* London: Heinemann, 1958.
Larguèche, Abdelhamid. "Nasîm Shammamâ: Un caïd face à lui-même et face aux autres." In *Juifs et Musulmans en Tunisie: Fraternité et déchirements*, ed. Sonia Fellous, 143–57. Paris: Somogy, 2003.
Laroui, Abdallah. *The History of the Maghrib: An Interpretive Essay.* Princeton, NJ: Princeton University Press, 1977.
Le Gall, Michel F., and Kenneth J. Perkins, eds. *The Maghrib in Question: Essays in History and Historiography.* Austin: University of Texas Press, 1997.
Lemieux, Cyril. "À quoi sert l'analyse des controverses?" *Mil neuf cent. Revue d'histoire intellectuelle* 25, no. 1 (2007): 191–212.
Levi Della Vida, Giorgio. "David Santillana." *Rivista degli studi orientali* 12 (1929–30): 453–61.
Lewis, Mary Dewhurst. *Divided Rule: Sovereignty and Empire in French Tunisia, 1881–1938.* Berkeley: University of California Press, 2013.
——. "Geographies of Power: The Tunisian Civic Order, Jurisdictional Politics, and Imperial Rivalry in the Mediterranean, 1881–1935." *Journal of Modern History* 80, no. 4 (2008): 791–830.
Liauzu, Claude. "Guerre des Sabines et tabou du métissage: Les mariages mixtes de l'Algérie coloniale à l'immigration en France." *Les cahiers du CEDREF* 8–9 (2000). http://journals.openedition.org/cedref/207.

Loualich, Fatiha. "'Alim Families in Seventeenth-Century and Eighteenth-Century Algiers: Genealogy and Heritage." *History of the Family* 16, no. 2 (2011): 98–107.

Macken, Richard A. "The Indigenous Reaction to the French Protectorate in Tunisia, 1881–1900." Ph.D. diss., Princeton University 1973.

Maḥfūdh, Muḥammad. *Tarājim al-mu'allifin al-Tūnisiyyin*. Beyrouth: Dār al-Gharb al-Islāmī, 1982.

Mahjoubi, Ali. *L'Établissement du protectorat français en Tunisie*. Tunis: Université de Tunis, 1977.

Majma' al-Tūnisī lil-'Ulūm wa-al-Ādāb wa-al-Funūn. *Al-Shaykh al-Muṣliḥ Sālim Bū Ḥājib wa-ishkālīyat al-'aṣr:* Carthage: Bayt al-Ḥikmah, 2007.

Makdisi, Ussama. "Ottoman Orientalism." *American Historical Review* 107, no. 3 (2002): 768–96.

Maksud, Hamza Abdel. *The Public Debt of Egypt, 1854-1876*. Cairo: Government Press, 1944.

Mantran, Robert. *Inventaire des documents d'archives turcs du Dar El-Bey (Tunis)*. Paris: Presses universitaires de France, 1961.

Marglin, Jessica. "Between Tolerance and Persecution: North Africans on North African Jewish History." In *After Orientalism: Critical Perspectives on Western Agency and Eastern Re-Appropriations*, ed. François Pouillon and Jean-Claude Vatin, 64–73. Leiden: Brill, 2014.

——. "The Two Lives of Mas'ud Amoyal: Pseudo-Algerians in Morocco, 1830–1912." *International Journal of Middle East Studies* 44, no. 4 (2012): 651–70.

Marpeau, Benoit. *Gustave Le Bon: Parcours d'un intellectuel 1841-1931*. Paris: CNRS, 2000.

Martel, André. *Les Confins saharo-tripolitains de la Tunisie (1881-1911)*. Paris: Presses universitaires de France, 1965.

Matar, Nabil. *Islam in Britain, 1558-1685*. Cambridge: Cambridge University Press, 1998.

Mayeur-Jaouen, Catherine. "Les Familles au Moyen-Orient moderne et contemporain (XVIIe–XXIe siècle)." In *La Famille en islam: D'après les sources arabes*, 391–489. Paris: Les Indes savantes, 2013.

Mayeur-Jaouen, Catherine, ed. *Saints et héros du moyen-orient contemporain*. Paris: Maisonneuve et Larose, 2002.

Mbembé, Achile, and Libby Meintjes. "Necropolitics." *Public Culture* 15, no. 1 (2003): 11–40.

McDougall, James. "Crisis and Recovery Narratives in Maghrebi Histories of the Ottoman Period (ca. 1870–1970)." *Comparative Studies of South Asia, Africa, and the Middle East* 31, no. 1 (2011): 131–48.

——. *History and the Culture of Nationalism in Algeria*. Cambridge: Cambridge University Press, 2006.

——. *A History of Algeria*. Cambridge: Cambridge University Press, 2017.

——. "A World No Longer Shared: Losing the *droit de cité* in Nineteenth-Century Algiers." *Journal of Economic and Social History of the Orient* 60, no. 1–2 (2017): 18–49.

McGregor, Andrew. "The Tunisian Army in the Crimean War: A Military Mystery." *Military History Online*, December 22, 2006. https://www.aberfoylesecurity.com/?p=1195.

Meddeb, Abdelwahab, and Benjamin Stora, eds. *A History of Jewish-Muslim Relations: From the Origins to the Present Day*. Princeton, NJ: Princeton University Press, 2013.

Medici, Ana Maria. "Esclavage, Armée et Réformes à Tunis. Vue d'un des derniers mamelouks à la cour du Bey (XIXe siècle)." In *D'esclaves à soldats miliciens et soldats d'origine servile (XIIIe-XXIe siecles)*, ed. Carmen Bernand and Alessandro Stella, 75–82. Paris: L'Harmattan, Paris 2006.

—. "Européens à Tunis, Tunisiens en Europe. Croisée de regards sur la Méditerranée (1857–1877)." In *Europe, Its Borders and the Others*, ed. Luciano Tosi, 99–107. Naples: Edizioni scientifiche italiane, 2000.

Meeker, Michael E. *A Nation of Empire: The Ottoman Legacy of Turkish Modernity*. Berkeley: University of California Press, 2002.

Melfa, Daniela. *Les Apports culturels des réfugiés politiques italiens en Tunisie au XIXe siècle*. Rome: Aracne, 2008.

—. "À la recherche des voix des 'subalternes': Le cas de Thibar vu à travers les archives des Missionnaires d'Afrique." In *Chantiers et défis de la recherche sur le Maghrib contemporain*, ed. Pierre-Robert Baduel, 135–58. Paris: Karthala, 2009.

—. *Migrando a Sud: coloni italiani in Tunisia, 1881–1939*. Rome: Aracne, 2008.

Melikoff, Irène. "Le Prince Mustafa Fazil et l'intelligentsia ottomane." *Revue d'histoire maghrébine* 17 (1990): 193–99.

Melograni, Piero, and Lucetta Scaraffia, eds. *La Famiglia italiana dall'Ottocento a oggi*. Rome: Ed. Laterza, 1988.

Messick, Brinkley. *The Calligraphic State: Textual Domination and History in a Muslim Society*. Berkeley: University of California Press, 1993.

—. "Evidence: From Memory to Archive." *Islamic Law and Society* 9, no. 2 (2002): 231–70.

Meynier, Gilbert. *L'Algérie révélée: La guerre de 1914–1918 et le premier quart du* XXe *siècle*. Geneva: Droz, 1981.

Michel, Nicolas. *L'Égypte des villages autour du seizième siècle*. Louvain: Peeters, 2018.

Mills, Amy, James A. Reilly, and Christine Philliou. "The Ottoman Empire from Present to Past: Memory and Ideology in Turkey and the Arab World." *Comparative Studies of South Asia, Africa and the Middle East* 31, no. 1 (2011): 133–36.

Minawi, Mostafa. *The Ottoman Scramble for Africa: Empire and Diplomacy in the Sahara and the Hijaz*. Stanford, CA: Stanford University Press, 2016.

Miquel, André. *La Géographie humaine du monde musulman jusqu'au milieu du 11e siècle. Géographie et géographie humaine dans la littérature arabe des origines à 1050*. Paris: École des hautes études en sciences sociales, 1967.

Moalla, Asma. *The Regency of Tunis and the Ottoman Porte, 1777–1814: Army and Government of a North-African Ottoman Eyālet at the End of the Eighteenth Century*. London: Routledge, 2004.

Montalbano, Gabriele. "Les Italiens de Tunisie: La construction de l'italianité dans un contexte colonial français (1896–1918)." Ph.D. diss., École Pratique des Hautes Etudes, 2018.

Montana, Ismael Musah. *The Abolition of Slavery in Ottoman Tunisia*. Gainesville: University Press of Florida, 2013.

Moreau, Odile. "Une guerre 'sur les marges.' La Tunisie et la guerre de Crimée par-delà l'oubli." In *Réformes de l'Etat et réformismes au Maghreb (XIXe-XXe siècles)*, ed. Odile Moreau, 53–76. Paris: L'Harmattan-IRMC, 2009.

Mrad Dali, Inès. "De l'esclavage à la servitude: Le cas des Noirs de Tunisie." *Cahiers d'études africaines* 179–80 (2005): 935–55.

—. *Identités multiples et multitudes d'histoire: Les 'Noirs tunisiens' de 1846 à aujourd'hui.* Paris: EHESS, 2009.

Noureddine, Ali. "Opposants et serviteurs: Les deux faces de la magistrature française dans la Régence de Tunis (1883–1886)." In *Contributions du séminaire sur les administrations coloniales (2009-2010)*. Paris: IHTP, 2011.

Nouschi, André. *Enquête sur le niveau de vie des populations rurales constantinoises de la conquête jusqu'en 1919: Essai d'histoire économique et sociale*. Paris: Presses universitaires de France, 1961.

Ortner, Sherry B. "Resistance and the Problem of Ethnographic Refusal." *Comparative Studies in Society and History* 37, no. 1 (1995): 173–93.

Oualdi, M'hamed. *Esclaves et maîtres: Les mamelouks des beys de Tunis du XVIIe siècle aux années 1880*. Paris: Publications de la Sorbonne, 2011.

—. "L'Héritage du général Ḥusayn: La pertinence du 'national' et de la 'nationalité' au début du protectorat français sur la Tunisie." In *Penser le national au Maghreb et ailleurs*, ed. Fatma Ben Slimane and Hichem Abdessamad, 65–88. Tunis: Arabesques, 2012.

—. "À l'école des palais: Les maîtrises de l'écrit parmi les mamelouks des beys de Tunis, des années 1770 aux années 1860." *European Journal of Turkish Studies* 6 (2007). https://journals.openedition.org/ejts/1403.

—. "Mamluks in Ottoman Tunisia: A Category Connecting State and Social Forces." *International Journal of Middle East Studies* 48, no. 3 (2016): 473–90.

—. "La Nationalité dans le monde arabe des années 1830 aux années 1960: Négocier les appartenances et le droit." *Revue des mondes musulmans et de la Méditerranée* 137 (2015): 13–28.

—. "Provincializing and Forgetting Ottoman Administrative Legacies: Sons and Grandsons of Beys' Mamluks Facing French Administrators of Tunisia (1890s–1930s)." *Comparative Studies of South Asia, Africa and the Middle East* 34, no. 2 (2014): 418–31.

Parks, Richard C. *Medical Imperialism in French North Africa: Regenerating the Jewish Community of Colonial Tunis*. Lincoln: University of Nebraska Press, 2017.

Passeron, Jean-Claude, and Jacques Revel, eds. *Penser par cas*. Paris: Éditions de l'École des hautes études en sciences sociales, 2005.

Peirce, Leslie P. *The Imperial Harem. Women and Sovereignty in the Ottoman Empire*. New York: Oxford University Press, 1993.

Perkins, Kenneth J. *A History of Modern Tunisia*. New York: Cambridge University Press, 2004.

—. "'The Masses Look Ardently to Istanbul': Tunisia, Islam, and the Ottoman Empire, 1837–1931." In *Islamism and Secularism in North Africa*, ed. John Ruedy, 23–36. New York: Saint Martin's Press, 1994.

Perrier, Antoine. "La liberté des protégés Souverains, ministres et serviteurs des monarchies marocaine et tunisienne sous protectorat français (1881–1956)." Ph.D. diss., Sciences Po, 2019.

Philliou, Christine. *Biography of an Empire. Governing Ottomans in an Age of Revolution.* Berkeley: University of California Press, 2011.

Picard, Christophe, and Jacques Alexandropoulos. "'Orient' et 'Occident' dans l'historiographie du Maghreb de la période coloniale à l'époque actuelle." In *Décoloniser l'histoire? De l'histoire coloniale aux histoires nationales en Amérique latine et en Afrique, XIXe ^e–XXe siècles*, ed. Sophie Dulucq and Colette Zytnicki, 85–100. Saint-Denis: Société française d'histoire d'outre-mer, 2003.

Planas, Natividad. "Musulmans invisibles dans les territoires insulaires du roi d'Espagne." In *Les Musulmans dans l'histoire de l'Europe*, ed. Jocelyne Dakhlia and Bernard Vincent, 557–92. Paris: Albin Michel, 2011.

Pomian, Krzysztof. *Collectionneurs, amateurs et curieux: Paris, Venise, XVIe–XVIIIe siècle.* Paris: Gallimard, 1987.

Poncet, Jean. *La Colonisation et l'agriculture européennes en Tunisie depuis 1881, étude de géographie historique et économique.* Paris: Mouton, 1962.

Pouchepadass, Jacques. "Les 'Subaltern Studies' ou la critique postcoloniale de la modernité." *L'Homme* 156 (2000): 161–85.

Poumarède, Géraud. "Naissance d'une institution royale: Les consuls de la nation française en Levant et en Barbarie aux XVIe et XVIIe siècles." *Annuaire-Bulletin de la Société de l'histoire de France* (2001): 65–128.

Powers, David S. *The Development of Islamic Law and Society in the Maghrib: Qadis, Muftīs and Family Law.* Burlington, VT: Ashgate Variorum, 2011.

——. *Law, Society, and Culture in the Maghrib, 1300–1500.* Cambridge: Cambridge University Press, 2002.

Pravilova, Ekaterina A. *A Public Empire: Property and the Quest for the Common Good in Imperial Russia.* Princeton, NJ: Princeton University Press, 2014.

Qāsimī, Fatḥī. *Al-Shaykh Muḥammad Bayram al-Khāmis: Hayātuhu wa-Fikruhu al-Iṣlāḥī.* Carthage: Bayt al-Ḥikmah, 1990.

Quataert, Donald. *The Ottoman Empire, 1700–1922.* Cambridge: Cambridge University Press, 2005.

Rahem, Karim. *Le Sillage de la tribu: Imaginaires politiques et histoire en Algérie, 1843–1993.* Paris: Riveneuve, 2008.

Raouf Hamza, Hassine. *Communisme et nationalisme en Tunisie: De la "libération" à l'indépendance, 1943–1956.* Tunis: Université de Tunis, 1994.

Rayburn, John C. "Count Joseph Telfener and the New York, Texas, and Mexican Railway Company." *Southwestern Historical Quarterly* 68, no. 1 (1964): 29–42.

Raymond, André. "French Studies of the Ottoman Empire's Arab Provinces." *Mediterranean Historical Review* 19, no. 1 (2004): 54–72.

——. *Tunis sous les Mouradites:* La ville *et ses habitants au XVIIe siecle.* Tunis: Cérès éditions, 2006.

——. "Tunisiens et Maghrébins au Caire au XVIIIe siècle." *Cahiers de Tunisie* 26–27 (1959): 335–71.

Reeva, Simon. "The Education of an Iraqi Ottoman Army Office." In *The Origins of Arab Nationalism*, ed. Rashid Khalidi, 151–66. New York: Columbia University Press, 1991.

Reimer, Michael J. "Egyptian Views of Ottoman Rule: Five Historians and Their Works, 1820–1920." *Comparative Studies of South Asia, Africa and the Middle East* 31, no. 1 (2011): 149–63.

Renucci, Florence. "David Santillana, acteur et penseur des droits musulman et européen." *Monde(s)* 7, no. 1 (2015): 25–44.

Revault, Jacques. *Palais et demeures de Tunis: XVIIIe et XIXe siècles.* Paris: Editions du Centre national de la recherche scientifique, 1971.

Rivet, Daniel. "Le Fait colonial et nous. Histoire d'un éloignement." *Vingtième Siècle. Revue d'histoire* 33 (1992): 127–38.

——. *Le Maghreb à l'épreuve de la colonisation.* Paris: Hachette littératures, 2009.

Rossi, Benedetta. *From Slavery to Aid: Politics, Labour, and Ecology in the Nigerien Sahel, 1800–2000.* New York: Cambridge University Press, 2015.

Rothman, E. Natalie. *Brokering Empire: Trans-Imperial Subjects Between Venice and Istanbul.* Ithaca, NY: Cornell University Press, 2012.

Rothschild, Emma. *The Inner life of Empires an Eighteenth-Century History.* Princeton, NJ: Princeton University Press, 2011.

Ruedy, John. *Land Policy in Colonial Algeria.* Berkeley: University of California Press, 1967.

Saada, Emmanuelle. *Les Enfants de la colonie: Les métis de l'empire français entre sujétion et citoyenneté.* Paris: Découverte, 2007.

Sadok, Boubaker. *La Régence de Tunis au XVIIe siècle: Ses relations commerciales avec les ports de l'Europe méditerranéenne, Marseille et Livourne.* Zaghouan: Ceroma, 1987.

Safi, Polat. "History in the Trench: The Ottoman Special Organization—Teşkilat-ı Mahsusa Literature." *Middle Eastern Studies* 48, no. 1 (2012): 89–106.

——. *The Ottoman Special Organization—Teşkilat-ı Mahsusa. An Inquiry Into Its Operation and Administrative Characteristics.* Ankara: Bilkent University, 2012.

Sahli, Mohamed Chérif. *Décoloniser l'histoire, introduction à l'histoire du Maghreb.* Paris: F. Maspero, 1965.

Sapio, Luigi. "Sistema giuridico, contratto agrario e sviluppo rurale in Tunisia." In *Tunisia e Toscana*, ed. Vittorio Antonio Salvadorini, 353–94. Pisa: Edistudio, 2002.

Sartori, Paolo. "Premessa. Waqf, colonialismo e pluralismo giuridico nelle società musulmane." *Quaderni storici* 132 (2009): 627–52.

Sayad, Abdelmalek. "Elghorba: Le mécanisme de reproduction de l'émigration." *Actes de la recherche en sciences sociales* 1–2 (1975): 50–66.

Schacht, Joseph. *Introduction au droit musulman.* Paris: Maisonneuve et Larose, 1999.

Schaub, Jean-Frédéric. "La Catégorie 'études coloniales' est-elle indispensable?" *Annales. Histoire, Sciences Sociales* 63, no. 3 (2008): 625–46.

Schreier, Joshua. *The Merchants of Oran: A Jewish Port at the Dawn of Empire.* Stanford, CA: Stanford University Press, 2017.

Schroeter, Daniel. "Identity and Nation: Jewish Migrations and Intercommunity Relations in the Colonial Maghrib." In *La Bienvenue et l'adieu: Migrants juifs et musulmans au Maghreb, XVe–XXe siècles: Actes du colloque d'Essaouira migrations identité et modernité au Maghreb*, ed. Karima Dirèche and Frédéric Abécassis, 1:125–39. Casablanca: Karthala, 2012.

Scott, David. *Conscripts of Modernity. The Tragedy of Colonial Enlightenment.* Durham, NC: Duke University Press, 2004.

Scott, James C. *La Domination et les arts de la résistance. Fragments du discours subalterne.* Paris: Editions Amsterdam, 2008.
Sebag, Paul. *L'Evolution d'un ghetto nord-africain: la Ḥara de Tunis.* Paris: Presses universitaires de France, 1959.
——. *Tunis: Histoire d'une ville.* Paris: L'Harmattan, 1998.
Shorrock, William I. "The Tunisian Question in French Policy Toward Italy, 1881–1940." *International Journal of African Historical Studies* 16, no. 4 (1983): 631–51.
Shuval, Tal. *La Ville d'Alger vers la fin du XVIIIe siècle: Population et cadre urbain.* Paris: CNRS, 1998.
Slama, Bice. *L'Insurrection de 1864 en Tunisie.* Tunis: Maison tunisienne de l'édition, 1967.
Smida, Mongi. *Consuls et consulats de Tunisie au 19e siècle.* Tunis: Imprimerie de l'Orient, 1991.
——. *Khereddine: Ministre réformateur, 1873–1877.* Tunis: Maison tunisienne de l'édition, 1971.
Souriau-Hoebrechts, Christiane. *La Presse maghrébine: Libye, Tunisie, Maroc, Algérie; évolution historique, situation en 1965, organisation et problèmes actuels.* Paris: Éditions du Centre national de la recherche scientifique, 1969.
Sraïeb, Nourredine. *Le Collège Sadiki de Tunis. 1875–1956. Enseignement et nationalisme.* Paris: CNRS, 1995
Stein, Sarah Abrevaya. "Protected Persons? The Baghdadi Jewish Diaspora, the British State, and the Persisitence of Empire." *American Historical Review* 116, no. 1 (2011): 80–108.
——. "Sephardi and Middle Eastern Jewries Since 1492." In *The Oxford Handbook of Jewish Studies*, ed. Martin Goodman, Jeremy Cohen, and David Jan Sorkin, 327–62. Oxford: Oxford University Press, 2005.
Stillman, Norman A. "Réflexions sur l'influence des tanzimat sur la condition sociale et juridique des Juifs d'Afrique du Nord au XIXe siècle." In *Les Relations entre juifs et musulmans en Afrique du Nord, XIXe–XXe siècles*, 49–55. Paris: Éditions du Centre national de la recherche scientifique, 1980.
Stoler, Ann Laura. *Along the Archival Grain. Epistemic Anxieties and Colonial Common Sense.* Princeton, NJ: Princeton University Press, 2009
——. "Tense and Tender Ties: The Politics of Comparison in North American History and (Post) Colonial Studies." *Journal of American History* 88, no. 3 (2001): 829–65.
Stora, Benjamin. "The Crémieux Decree." In *A History of Jewish-Muslim Relations: From the Origins to the Present Day*, ed. Abdelwahab Meddeb and Benjamin Stora, trans. Jane Marie Todd and Michael B. Smith, 286–91. Princeton, NJ: Princeton University Press, 2013.
——. *Messali Hadj, 1898–1974.* Paris: Sycomore, 1982.
——. *Les Trois Exils: Juifs d'Algérie.* Paris: Stock, 2006.
Streiff-Fenart, Jocelyne. "Le 'Métissage' franco-algérien. Catégories politiques et imaginaires à propos des mariages mixtes." *Annuaire de l'Afrique du Nord* 29 (1990): 343–51.
Subrahmanyam, Sanjay. "Beyond Incommensurability: Understanding Inter-Imperial Dynamics." *Department of Sociology, UCLA Theory and Research in Comparative Social Analysis* 32 (2005): 1–38.

Surun, Isabelle, ed. *Les Sociétés coloniales à l'âge des Empires: 1850–1960*. Neuilly: Atlande, 2012.

Taïeb, Jacques. "Les Juifs livournais et la modernité occidentale." In *Entre Orient et Occident: Juifs et musulmans en Tunisie*, ed. Denis Cohen-Tannoudji, 217–27. Paris, Éclat, 2007.

Taraud, Christelle. "Les Femmes, le genre et les sexualités dans le Maghreb colonial (1830–1962)." *CLIO: Femmes, Genre, Histoire* 33 (2011): 157–91.

Ṭawīlī, Aḥmad. *Al-Jinirāl Ḥusayn: ḥayātuhu wa-athāruhu*. Tunis: Baladīyat Tūnis, 1994.

Temimi, Abdeljelil. "The Arab-Ottoman Historical Heritage and Its Impact on the Arab-Turkish Relations." *Arab Historical Review for Ottoman Studies* 7–8 (1993): 215–25.

——. "Min aʿlāminā al-bārizīn wa al-mansiyyin, al-shaykh Ṣāliḥ al-Sharīf al-Tūnisī." *Revue d'histoire maghrébine* 23–24 (1981): 345–62.

——. *Recherches et documents d'histoire maghrébine: la Tunisie, l'Algérie et la Tripolitaine de 1816 à 1871*. Tunis: Université de Tunis, 1971.

Terem, Etty. *Old Texts, New Practices: Islamic Reform in Modern Morocco*. Stanford, CA: Stanford University Press, 2014.

Thénault, Sylvie. "L'État colonial. Une question de domination." In *Les Empires coloniaux: XIXe–XXe siècle*, ed. Pierre Singaravelou, 215–56. Paris: Points, 2013.

Thompson, Elizabeth. *Colonial Citizens: Republican Rights, Paternal Privilege, and Gender in French Syria and Lebanon*. New York: Columbia University Press, 2000.

Tobi, Joseph and Zivia. *Judeo-Arabic literature in Tunisia, 1850–1950*. Detroit, MI: Wayne State University Press, 2014.

Toledano, Ehud R. "The Emergence of Ottoman-Local Elites (1700–1900): A Framework for Research." In *Middle Eastern Politics and Ideas: A History from Within*, ed. Ilan Pappé and Moshe Ma'oz, 145–62. London: Tauris, 1997.

——. "Shemsigül: A Circassian Slave Mid-Nineteenth-Century Cairo." In *Struggle and Survival in the Modern Middle East*, ed. Edmunde Burke III, 59–74. New York: I. B. Tauris, 1993.

——. *Slavery and Abolition in the Ottoman Middle East*. Seattle: University of Washington Press, 1998.

Torre, Angelo. " 'Faire communauté:' Confréries et localité dans une vallée du Piémont (XVIIe–XVIIIe siècle)." *Annales. Histoire, Sciences Sociales* 62, no. 1 (2007): 101–35.

Triaud, Jean-Louis. "La Libye." In *Les Voies d'Allah*, ed. Alexandre Popovic and Gilles Veinstein, 409–12. Paris: Fayard, 1996.

Trivellato, Francesca. *The Familiarity of Strangers. The Sephardic Diaspora, Livorno, and Cross-Cultural Trade in the Early Modern Period*. New Haven, CT: Yale University Press, 2009.

——. "Is There a Future for Italian Microhistory in the Age of Global History?" *California Italian Studies* 2, no. 1 (2011): 1–24.

Tsur, Yaron. "Réformistes musulmans et juifs en Tunisie à la veille de l'occupation française." In *Juifs et musulmans en Tunisie. Fraternité et déchirements*, ed. Sonia Fellous, 161–69. Paris: Somogy, 2003.

Tunger-Zanetti, Andreas. *La Communication entre Tunis et Istanbul, 1860–1913: Province et métropole*. Paris: L'Harmattan, 1996.

Vaḥḥah, ʿAbd al-Qādir. *Al-Ḥaraka al-iṣlāḥīya al-zaytūnīya wa-al-madrasīya al-Tūnisīya: al-Shaykh Sālim Bū Ḥājib wa-manhajuhu al-iṣlāḥī, 1244–1342/1827–1828–1924*. Tunis: al-Ṭabʿah 1, 2007.

Valensi, Lucette. *Fellahs tunisiens: L'économie rurale et la vie des campagnes aux 18e et 19e siècles*. Paris: Mouton, 1977.

——. "Une histoire des Juifs de Tunisie est-elle nécessaire? Est-elle posible?" In *Histoire communautaire, histoire plurielle: La communauté juive de Tunsie*, ed. Habib Kazdaghli, 51–63. Tunis: CPU, 1999.

——. *Le Maghreb avant la prise d'Alger, 1790–1830*. Paris: Flammarion, 1969.

Van Krieken, Gerard. *Khayr al-Dīn et la Tunisie, 1850–1881*. Leiden: Brill, 1976.

Van Staëvel, Jean Pierre. "Le Qâdî au bout du labyrinthe: L'impasse dans la littérature jurisprudentielle mâlikite (al-Andalus et Maghreb, 3h/IXjc–9h/XVjc siècles)." In *L'Urbanisme dans l'Occident musulman au Moyen Âge. Aspects juridiques*, ed. Patrice Cressier, Maribel Fierro, and Jean Pierre Van Staëvel, 39–63. Madrid: Casa de Velázquez, 2000.

Vatin, Nicolas. "L'Inhumation intra-muros à Istanbul à l'époque ottoman." In *Les Ottomans et la mort: Permanences et mutation*, ed. Gilles Veinstein and Nicolas Vatin, 157–74. Leiden: Brill, 1996.

Vatin, Nicolas, and Gilles Veinstein. "Les Obsèques des sultans ottomans de Mehmed II à Ahmed 1er (1481–1616)." In *Les Ottomans et la mort: Permanences et mutations*, ed. Gilles Veinstein and Nicolas Vatin, 187–205. Leiden: Brill, 1996.

Vermeren, Hugo. *Les Italiens à Bône. Migrations méditerranéennes et colonisation de peuplement en Algérie (1865–1940)*. Paris: Nanterre University, 2015.

Vermeren, Pierre. *Misère de l'historiographie du Maghreb post-colonial: 1962–2012*. Paris: Publications de la Sorbonne, 2012.

Warscheid, Ismail. *Traduire le social en normatif: La justice islamique dans le grand Touat (Sahara algérien) au XVIIIe siècle*. Paris: EHESS, 2014.

Watenpaugh, Keith David. *Being Modern in the Middle East: Revolution, Nationalism, Colonialism, and the Arab Middle Class*. Princeton, NJ: Princeton University Press, 2006.

Weil, Patrick. *Qu'est-ce qu'un Français? Histoire de la nationalité française depuis la Révolution*. Paris: Grasset, 2002.

Wick, Alexis. "Sailing the Modern Episteme: Al-Ṭahṭāwī on the Mediterranean." *Comparative Studies of South Asia, Africa and the Middle East* 34, no. 2 (2014): 405–17.

Windler, Christian. *La Diplomatie comme experience de l'autre: Consuls francais au Maghreb (1700–1840)*. Geneva: Droz, 2002.

Yacono, Xavier. *Histoire de la colonisation française*. Paris: Presses universitaires de France, 1973.

Yazbak, Mahmoud. "Nabulsi Ulama in the Late Ottoman Period, 1864–1914." *International Journal of Middle East Studies* 29, no. 1 (1997): 71–91.

Yazidi, Béchir. *La Politique coloniale et le domaine de l'État en Tunisie: De 1881 jusqu'à la crise des années 1930*. Tunis: Sahar, 2005.

Ze'evi, Dror. "Kul and Getting Cooler: The Dissolution of Elite Collective Identity and the Formation of Official Nationalism in the Ottoman Empire." *Mediterranean Historical Review* 11, no. 2 (1996): 177–95.

Zilfi, Madeline C. *The Politics of Piety: The Ottoman Ulema in the Postclassical Age (1600–1800)*. Minneapolis, MN: Bibliotheca Islamica, 1988.

Zimmermann, Francis. "Sahlins, Obeyesekere et la mort du capitaine Cook." *L'Homme* 146 (1998): 191–205.
Zmerli, Sadok. *Les Précurseurs*. Tunis: Bouslama, 1964.
Zouari, Ali. *Les Relations commerciales entre Sfax et le Levant aux XVIII*[e] *et XIX*[e] *siècles*. Tunis: Institut national d'archéologie et d'art, 1990.
Zytnicki, Colette. "The 'Oriental Jews' of the Maghreb: Reinventing the North African Jewish Past in the Colonial Era." In *Colonialism and the Jews*, ed. Ethan B. Katz, Lisa Moses Leff, and Maud S. Mandel, 29–53. Bloomington: Indiana University Press, 2017.

Index

Page numbers in *italics* represent illustrations.

GPSR Authorized Representative: Easy Access System Europe, Mustamäe tee 50, 10621 Tallinn, Estonia, gpsr.requests@easproject.com

www.ingramcontent.com/pod-product-compliance
Ingram Content Group UK Ltd.
Pitfield, Milton Keynes, MK11 3LW, UK
UKHW040245010926
474715UK00005BA/17/J
9780231191869